Excel® 2007 Guide

S0-AIG-240

Understanding Basic Statistics

FIFTH EDITION

Charles Henry Brase

Regis University

Corrinne Pellillo Brase

Arapahoe Community College

Prepared by

John Stokes

Edutorial

BROOKS/COLE
CENGAGE Learning™

Australia • Brazil • Japan • Korea • Mexico • Singapore • Spain • United Kingdom • United States

© 2010 Brooks/Cole, Cengage Learning

ALL RIGHTS RESERVED. No part of this work covered by the copyright herein may be reproduced, transmitted, stored, or used in any form or by any means graphic, electronic, or mechanical, including but not limited to photocopying, recording, scanning, digitizing, taping, Web distribution, information networks, or information storage and retrieval systems, except as permitted under Section 107 or 108 of the 1976 United States Copyright Act, without the prior written permission of the publisher.

For product information and technology assistance, contact us at
**Cengage Learning Customer & Sales Support,
1-800-354-9706**

For permission to use material from this text or product, submit all requests online at **www.cengage.com/permissions**
Further permissions questions can be emailed to
permissionrequest@cengage.com

ISBN-13: 978-0-547-18916-1
ISBN-10: 0-547-18916-8

Brooks/Cole
25 Thomson Place
Boston, MA 02210
USA

Cengage Learning is a leading provider of customized learning solutions with office locations around the globe, including Singapore, the United Kingdom, Australia, Mexico, Brazil, and Japan. Locate your local office at:
www.cengage.com/international

Cengage Learning products are represented in Canada by Nelson Education, Ltd.

To learn more about Brooks/Cole, visit
www.cengage.com/brookscole

Purchase any of our products at your local college store or at our preferred online store
www.ichapters.com

Printed in Canada
1 2 3 4 5 6 7 12 11 10 09 08

Table of Contents

Copyright © Cengage Learning. All rights reserved.

Copyright © Cengage Learning. All rights reserved.

Preface

The use of computing technology can greatly enhance a student's learning experience in statistics. *Understanding Basic Statistics* is accompanied by four Technology Guides, which provide basic instruction, examples, and lab activities for four different tools:

TI-83 Plus and TI-84 Plus

Microsoft Excel® 2007 with Analysis ToolPak for Windows®

MINITAB Version 15

SPSS Version 16

The TI-83 Plus and TI-84 Plus are versatile, widely available graphing calculators made by Texas Instruments. The calculator guide shows how to use their statistical functions, including plotting capabilities.

Excel is an all-purpose spreadsheet software package. The Excel guide shows how to use Excel's built-in statistical functions and how to produce some useful graphs. Excel is not designed to be a complete statistical software package. In many cases, macros can be created to produce special graphs, such as box-and-whisker plots. However, this guide only shows how to use the existing, built-in features. In most cases, the operations omitted from Excel are easily carried out on an ordinary calculator. The Analysis ToolPak is part of Excel and can be installed from the same source as the basic Excel program (normally, a CD-ROM) as an option on the installer program's list of Add-Ins. Details for getting started with the Analysis ToolPak are in Chapter 1 of this guide. No additional software is required to use the Excel functions described.

MINITAB is a statistics software package suitable for solving problems. It can be packaged with the text. Contact Cengage Learning for details regarding price and platform options.

SPSS is a powerful tool that can perform many statistical procedures. The SPSS guide shows how the manage data and perform various statistical procedures using this software.

The lab activities that follow accompany the text *Understanding Basic Statistics,* 5th edition by Brase and Brase. On the following page is a table to coordinate this guide with the parent text *Understandable Statistics,* 9th edition by Brase and Brase.

In addition, over one hundred data files from referenced sources are described in the Appendix. These data files are available via download at:

www.cengage.com/statistics/Brase/UBS5e

Copyright © Cengage Learning. All rights reserved.

Understanding the Differences Between *Understandable Statistics* 9/e and *Understanding Basic Statistics* 5/e

Understandable Statistics is the full, two-semester introductory statistics textbook, which is now in its Ninth Edition.

Understanding Basic Statistics is the brief, one-semester version of the larger book. It is currently in its Fifth Edition.

Unlike other brief texts, *Understanding Basic Statistics* is not just the first six or seven chapters of the full text. Rather, topic coverage has been shortened in many cases and rearranged, so that the essential statistics concepts can be taught in one semester.

The major difference between the two tables of contents is that Regression and Correlation are covered much earlier in the brief textbook. In the full text, these topics are covered in Chapter 10. In the brief text, they are covered in Chapter 4.

Analysis of a Variance (ANOVA) is not covered in the brief text.

Understanding Statistics has 12 chapters and *Understanding Basic Statistics* has 11. The full text is a hardcover book, while the brief is softcover.

The same pedagogical elements are used throughout both texts.

The same supplements package is shared by both texts.

Following are the two Tables of Contents, side-by-side:

	Understandable Statistics (full)	*Understanding Basic Statistics* (brief)
Chapter 1	Getting Started	Getting Started
Chapter 2	Organizing Data	Organizing Data
Chapter 3	Averages and Variation	Averages and Variation
Chapter 4	Elementary Probability Theory	Correlation and Regression
Chapter 5	The Binomial Probability Distribution and Related Topics	Elementary Probability Theory
Chapter 6	Normal Distributions	The Binomial Probability Distribution and Related Topics
Chapter 7	Introduction to Sampling Distributions	Normal Curves and Sampling Distributions
Chapter 8	Estimation	Estimation
Chapter 9	Hypothesis Testing	Hypothesis Testing
Chapter 10	Correlation and Regression	Inferences About Differences
Chapter 11	Chi-Square and F Distributions	Additional Topics Using Inference
Chapter 12	Nonparametric Statistics	

Copyright © Cengage Learning. All rights reserved.

Excel 2007 Guide

CHAPTER 1: GETTING STARTED

GETTING STARTED WITH EXCEL

Microsoft Excel® is an all-purpose spreadsheet application with many functions. We will be using Excel 2007. This guide is not a general Excel manual, but it will show you how to use many of Excel's built-in statistical functions. You may need to install the Analysis ToolPak from the original Excel software if your computer does not have it. To determine if your installation of Excel includes the Analysis ToolPak, open Excel, click on the **Office** button, and then click on the **Excel Options** button to open the Excel Options dialog box. In the Excel Options dialog box, select **Add-Ins.** If Analysis ToolPak doesn't appear under the Active Application Add-ins, select **Analysis ToolPak** from the **Inactive Application Add-ins** list and click **Go.** Check the **Analysis ToolPak** box and click **OK** in the Add-Ins dialog box to install the Analysis ToolPak.

If you are familiar with Windows-based programs, you will find that many of the editing, formatting, and file-handling procedures of Excel are similar to those you have used before. You use the mouse to select, drag, click, and double-click as you would in any other Windows program. If you have any questions about Excel not answered in this guide, consult the Excel manual or click on the **Excel Help** question mark in the upper right portion of the Ribbon.

The Excel Window

When you have opened Excel, you should see a window like this:

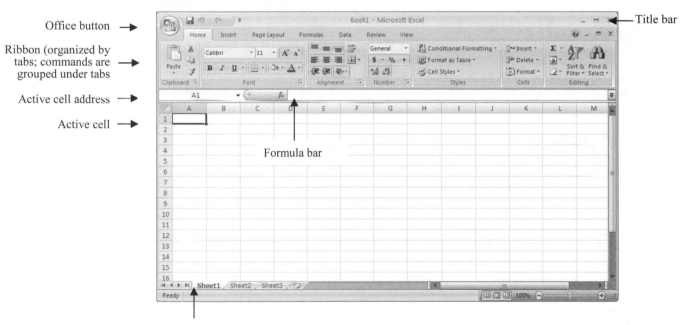

Office button →
Ribbon (organized by tabs; commands are grouped under tabs →
Active cell address →
Active cell →

Formula bar

← Title bar

Selected worksheet tab

The Excel Workbook

An Excel file is called a Workbook. Notice that in the display shown above, the Title bar shows Book 1 - Microsoft Excel. This means that we are working in Book 1.

Each workbook consists of one or more worksheets. In the worksheet above, the tabs near the bottom of the screen show that we are working with Sheet 1. To change worksheets, click on the appropriate tab. Alternatively, you can right-click the arrows just to the left of the worksheet tabs to get a list of all the worksheets in the workbook, and then select a worksheet.

Copyright © Cengage Learning. All rights reserved.

The Cells in the Worksheet

When you look at a worksheet, you should notice horizontal and vertical grid lines. If they are missing, you will need to activate that feature. To do so,

1. Select the **View tab** at the top of the Ribbon. Be sure that the **Gridlines** option in the **Show/Hide** group is checked.

2. Select the **Page Layout** tab. Be sure that the **View** box under **Gridlines** in the **Sheet Options** group is checked.

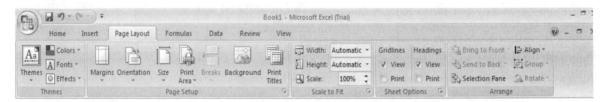

Cell Addresses

The cells are formed by intersecting rows and columns. A cell's address consists of a column letter followed by a row number. For example, the address B3 designates the cell that lies in Column B, Row 3. When Cell B3 is highlighted, it is the active cell. This means we can enter text or numbers into Cell B3.

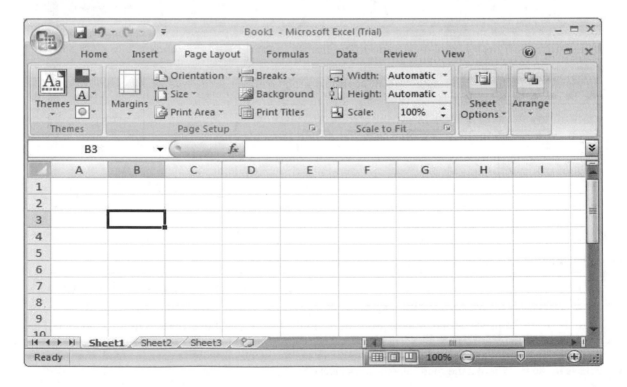

Copyright © Cengage Learning. All rights reserved.

Selecting Cells

To select a cell, position the cursor in the cell and click the left mouse button.

Sometimes you will want to select several cells at once, in order to format them (as described next). To select a rectangular block of cells, position the cursor in a corner cell of the block, hold down the left mouse button, drag the cursor to the cell in the block's opposite corner, and release the button. The selected cells will be highlighted, as shown below.

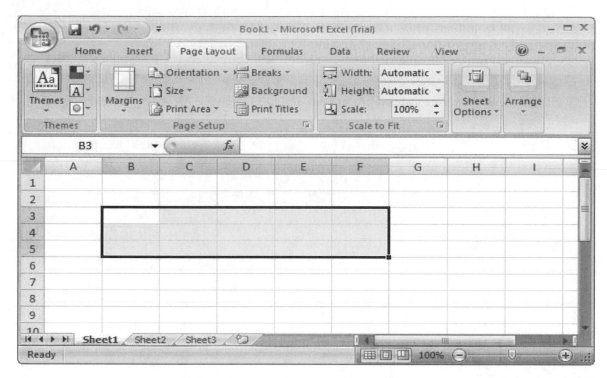

To select an entire column, click on the letter above it; to select an entire row, click on the number to its left. To select every cell in the worksheet, click on the blue, blank rectangle with the triangle in the lower right corner. This rectangle lies in the upper left corner of the worksheet above row header 1 and left of column header A.

You can also select a block of cells by typing the two corner cells into the active cell address window. The block highlighted in the figure above would be selected by typing B3:F5 and pressing Enter.

Formatting Cell Contents

In Excel, you may place text or numbers in a cell. As in other Windows 2007 applications, you can format the text or numbers by using the buttons for bold (**B**), italics (*I*), underline (U), etc, that are in the **Font** group under the **Home** tab. Other options include left, right, and centered alignment within a cell. These may be found in the **Alignment** group under the **Home** tab.

Numbers can be formatted to represent dollar amounts ($) or percents (%) and can be shown with commas in large numbers (,). The number of decimal places to which numbers are carried is also adjustable. All these options appear in the **Number** group under the **Home** tab. Other options are accessible in the **Cells** group under the **Home** tab.

Changing Cell Width

To change the column widths and row heights for selected cells, select the **Format** option in the **Cells** group under the **Home** tab. The commands for changing cell size are listed under the **Cell Size** heading.

Copyright © Cengage Learning. All rights reserved.

Column widths and row heights can also be adjusted by placing the cursor between two column letters or row numbers. When the cursor changes appearance, hold down the left mouse button, move the column or row boundary, and release.

All these instructions may seem a little mysterious. Once you try them, however, you will find that they are fairly easy to remember.

ENTERING DATA

In Excel we enter data and labels in the cells. It is common to select a column for the data and place a label as the first entry in the column.

Let's enter some data on television advertising. For each of twenty hours of prime-time viewing, both the number of ads and the time devoted to ads were recorded. We will enter the data in two columns, as shown:

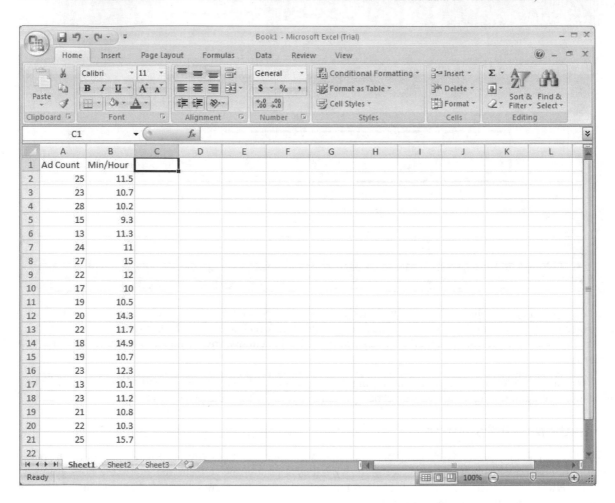

Entering and Correcting Data

To select a cell for content entry, move the mouse pointer to the cell and click. Then type the label or data and press Enter. Excel automatically moves to select the next cell in the same column. If you want to enter information in a different cell, just click on it.

Copyright © Cengage Learning. All rights reserved.

Errors are easily fixed. If you notice a mistake before you press Enter, simply back-arrow to the mistake and correct it. If you notice the error after you have pressed Enter, select the affected cell and then click on the Formula bar to add a typing cursor to the cell contents displayed. Use standard keyboard editing techniques to make corrections, then press Enter.

If you want to erase the contents of a cell or range of cells but keep the formatting, select the (Clear) button in the **Editing** group under the **Home** tab. This will produce a menu with four options. Select the **Clear Contents** option. Other options under **Clear** have slightly different effects. The **Clear Formats** option keeps the content but clears the format. The **Clear All** option clears both content and format. **Clear Formats** is especially useful for changing percent data back into decimal format.

Arithmetic Options under the Home Tab

Summing Data in a Column

In the **Editing** group under the **Home** tab, the (Sum) button automatically sums the values in the selected cells. When we sum the contents of an entire column, Excel places the sum under the selected cells. It is a good idea to type the label "Total" next to the cell where the total appears. Below, we selected cells in Column A containing numerical values (A2:A21), pressed the button, and then typed the word "Total" in the corresponding row of Column C. We see that the total of Column A is 419.

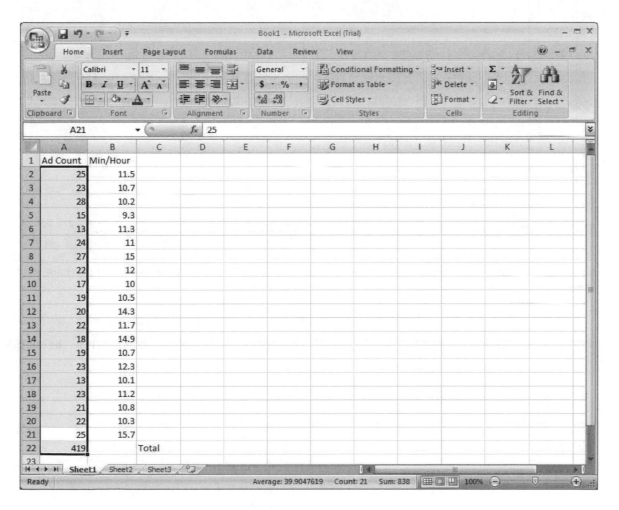

Copyright © Cengage Learning. All rights reserved.

Sorting Data

Two commands are available to sort data in ascending and descending order: <u>**Sort**</u> **A to Z** (or **Sort Smallest to Largest**), and <u>**Sort**</u> **Z to A** (or **Sort Largest to Smallest**, respectively). To sort just one column, highlight that column, click the **Sort & Filter** button in the **Editing** group under the **Home** tab, and select the desired sorting option from the drop-down menu. To sort two or more columns by ascending or descending order of the data in the first column, highlight all the columns and select the appropriate option. In general, we will simply sort one column of data at a time, as shown.

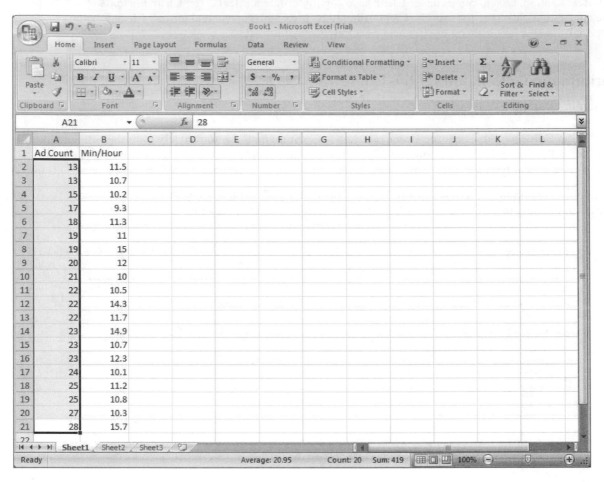

Notice that the data in the first column are now in ascending order. The data in the second column have not moved.

If you decide that it was a mistake to sort the data this way, and you have not made any other changes since you did the sorting, you can use the (**Undo**) button in the **Quick Access** toolbar, which is on the left side of the Title bar. The data will appear again in their original order.

Copyright © Cengage Learning. All rights reserved.

Copying Cells

To copy one cell or a block of cells to another location on the worksheet,

(a) Select the cells you wish to copy

(b) Click the ▣ (**Copy**) button in the **Clipboard** group under the **Home** tab. (The shortcut for this process is **Ctrl-C**.) Notice that the range of cells being copied now has a blinking border around it.

(c) Select the upper-left cell of the block that will receive the copy.

(d) Press Enter. When you press Enter, the copy process is complete and the blinking border around the original cells disappears.
Note: Even if you use the **Paste** command in the **Clipboard** group under the **Home** tab, or the shortcut **Ctrl-V** to paste, you must still press Enter to remove the blinking border around the original cells.

To copy one cell or a range of cells to another worksheet or workbook, follow steps (a) and (b) above. For step (c), be sure you are in the destination worksheet or workbook and that the worksheet or workbook is activated. Then proceed to step (d).

USING FORMULAS

A formula is an expression that generates a numerical value in a cell, usually based on values in other cells. Formulas usually involve standard arithmetic operations. Excel uses + for addition, - (hyphen) for subtraction, * for multiplication, / for division, and ^ (carat) for exponentiation (raising to a power).

For instance, if we want to divide the contents of Cell A2 by the contents of Cell B2 and place the results in Cell C2, we do the following:

(a) Make Cell C2 the active cell.

(b) Click in the Formula bar and type =A2/B2.

(c) Press Enter.

The value in Cell C2 will be the quotient of the values in Cells A2 and B2.

If, for a whole series of rows, we wanted to divide the entry in Column A by the entry in Column B and put the results in Column C, we could repeat the above process over and over. However, the typing would be tedious. We can accomplish the same thing more easily by copying and pasting:

(a) Type =A2/B2 in Cell C2 as described above.

(b) Move the cursor to the lower right corner of Cell C2. The cursor should change shape to small black cross (+). Now hold down the left mouse button and drag the + down until all the cells in Column C in which you want the calculation done are highlighted.

(c) Release the mouse button. The cell entries in Column C should equal the quotients of the same-row entries in Columns A and B.

Copyright © Cengage Learning. All rights reserved.

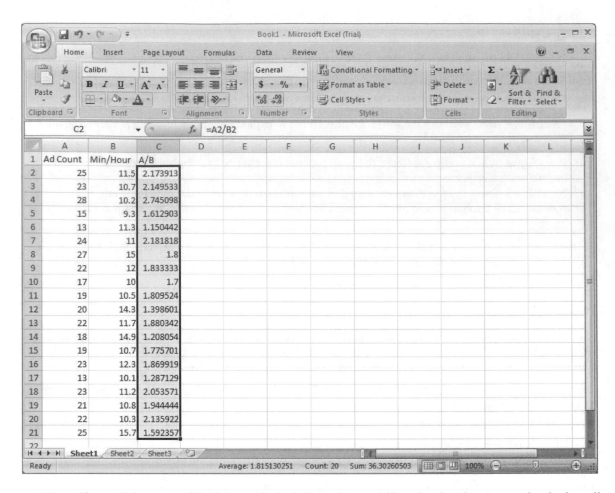

Now, if you click on one of the lower cells in Column C, you will notice that the row number in the cell addresses is not 2, but rather the number of the new cell's row. For instance, if the formula =A1+B3 is inserted in Column C for data in Columns A and B, the formula for Cells C1, C2, and C3 would be =A1+B3, =A2+B4, and =A3+B5, respectively.

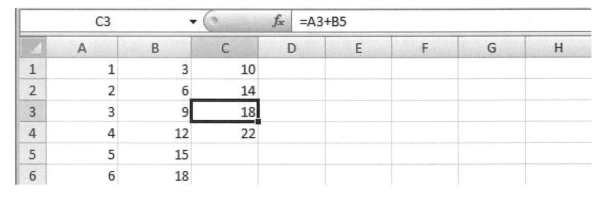

Similarly, when the value in Cell C1 is copied to Cell D2, the formula for Cell D2 is =B2+C4, and so the value in that window equals the sum of the Cells B2 and C4.

Copyright © Cengage Learning. All rights reserved.

D2			f_x	=B2+C4			
A	B	C	D	E	F	G	H
1	3	10					
2	6	14	28				
3	9	18					
4	12	22					
5	15						
6	18						

Sometimes you will want to prevent the automatic address adjustment. To do this put a dollar sign before any row or column number you want to keep from changing. For example, when the formula =A$1+$B3 is copied to Cell D2, the formula changes to =B$1+$B4. Placing a dollar sign in front of the "1" keeps this row fixed in the formula. Similarly, the dollar sign in front of the "B" keeps this column fixed in the formula, regardless of the cell to which the formula is copied.

D2			f_x	=B$1+$B4			
A	B	C	D	E	F	G	H
1	3	10					
2	6	14	15				
3	9	18					
4	12	22					
5	15						
6	18						

We will call an address with two $ signs (e.g., "$C$5") an *absolute* address, because it always refers to the same cell (C5), no matter where the formula is/pasted. A cell with only one $ sign in it, or none at all, is called a *relative* address, because the column and/or row of the cell referred to can change as the formula is pasted from one location to another.

SAVING WORKBOOKS

After you have entered data into an Excel spreadsheet, it is a good idea to save it. Click on the **Office** button and select **Save As.** Select the **Excel Workbook** option. A dialog box like the one on the next page will appear.

Copyright © Cengage Learning. All rights reserved.

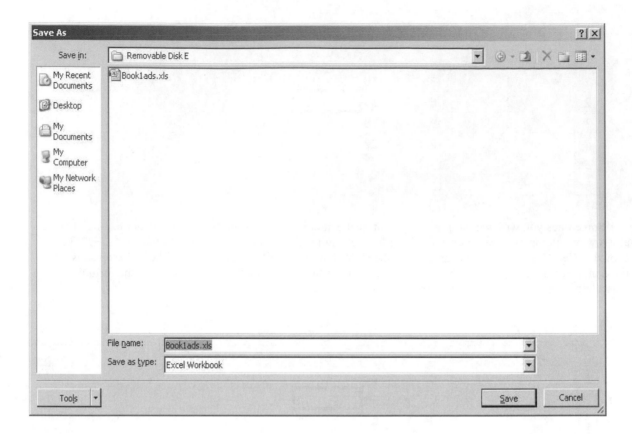

If you are in a college computer lab, you might save your files to a removable media device. We named the workbook on TV ads as Book1ads.

It is a good idea to save your workbook periodically as you are working on it. After you have saved the workbook for the first time, you can save updates during your working session by using the **Save** button on the **Quick Access** toolbar, on the left side of the Title bar. **Save** is the first button to the right of the **Office** button; it looks like a diskette.

To retrieve an Excel workbook, click on the **Office** button and select **Open.** Then select the desired workbook from the **Recent Documents** list.

Printing Your Worksheets

Clicking the **Office** button and selecting **Print,** and then choosing **Print** under the **Preview and print the document** list will open the Print dialog box.

Copyright © Cengage Learning. All rights reserved.

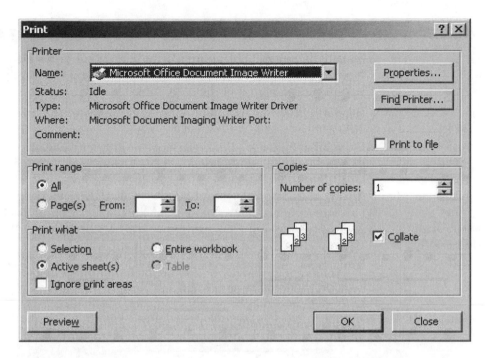

If you select a range of cells on the worksheet before you print, you may print the selected material. Notice that you can tell the printer what to print by clicking next to **Selection, Active sheet(s)** or **Entire workbook.** Clicking the **Preview** button lets you see what you will print before the page is printed.

LAB ACTIVITIES FOR GETTING STARTED

1. Go to your computer lab (or use your own computer) and open Excel. Check to see if you have the Analysis ToolPak add-in. If so, be sure it is activated.

2. If you have not already done so, enter the TV Ad Count and Min/Hour data into the workbook. Use Column A for the Ad Count and Column B for the Min/Hour data.

3. Save the workbook as Book1ads.

4. Select the cells containing the labels and data and print.

5. **(a)** In Column C, place the quotients A/B for the Rows 2 through 21. Use the Formula bar in Cell C2, and drag downward with the little + symbol to complete the quotients for all rows. Note that if the calculation mode (listed by clicking the **Calculation Options** button in the **Calculation** group under the **Formulas** tab) has been set to **Manual,** you may need to use the key combination **Shift-F9** after the cells are highlighted.

    (b) Use the Σ (**Sum**) button in the **Editing** group under the **Home** tab to total the Ad Count column and to total the Min/Hour column. Place a label in Column D adjacent to the totals.

    (c) Select the data in Columns A, B, C, and D and print it.

6. In this problem we will copy a column of data and sort the copy.

    (a) Select Column A (Ad Count) and copy it to Column D.

    (b) Select Column D and sort it in ascending order (use only the original data, not the sum).

    (c) Select both Column A (Ad Count) and Column B (Min/Hour). Sort these columns in descending order. Are the data entries of 13 in Column A still next to the data entries 11.3 and 10.1 in Column B? Are the data in Column B sorted?

Copyright © Cengage Learning. All rights reserved.

RANDOM SAMPLES (SECTION 1.2 OF *UNDERSTANDING BASIC STATISTICS*)

Excel has several random number generators. The one we will find most convenient is the function **RANDBETWEEN(bottom,top)**. This function generates a random integer between (inclusively) whatever integer is put in for "bottom" and whatever integer is put in for "top."

To use RANDBETWEEN, select a cell in the active worksheet. In the Formula bar, type an equals sign and click the **Insert Function** button on the left side of the Formula bar:

Pressing this button calls up the Insert Function dialog box. Select All in the **Or select a category** window, and then scroll down in the **Select a function** window until you reach RANDBETWEEN. Note that this command is present only if you have the Analysis ToolPak checked under the **Active Application Add-ins** list in the **Excel Options Add-Ins** feature.

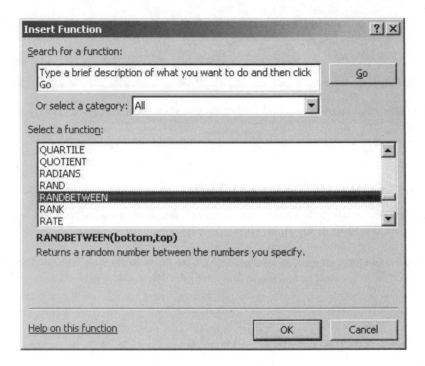

Select RANDBETWEEN and click **OK.** In the resulting Function Arguments dialog box, fill in the bottom and top numbers and click **OK.** Alternatively, you may simply type =RANDBETWEEN(bottom,top) in the Formula bar, with numbers in place of bottom and top.

The random number generators of Excel have the characteristic that whenever a command is entered anywhere in the active workbook, the random numbers change because they are recalculated. To prevent this from happening, change the recalculate mode from automatic to manual. Click the **Calculation Options** button in the **Calculation** group under the **Formulas** tab, and select **Manual.**

With automatic recalculation disabled, you can still recalculate by pressing the **Shift-F9** key combination. Let us see this in an example, where we select a list of random numbers in a designated range and sort the list in ascending order.

Copyright © Cengage Learning. All rights reserved.

Example

There are 175 students enrolled in a large section of introductory statistics. Draw a random sample of fifteen of the students.

We assign each of the students a distinct number between 1 and 175. To find the numbers of the fifteen students to be included in the sample, we do the following steps.

(a) Change the calculation mode to **Manual.**

(b) Type the label "Sample" in Cell A1.

(c) Select Cell A2.

(d) Type =RANDBETWEEN(1,175) in the Formula bar and press Enter.

(e) Position the mouse pointer in the lower right corner of Cell A2 until it becomes a + sign, and click-drag downward until you reach Cell A16. Release the mouse button. Then press **Shift-F9**.

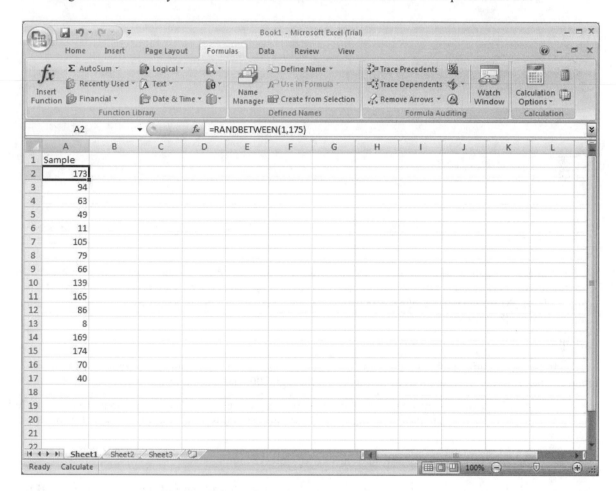

(f) To check for repetitions, use the **Sort & Filter** button in the **Editing** group under the **Home** tab. If there are repetitions, press **Shift-F9** again and re-sort.

Copyright © Cengage Learning. All rights reserved.

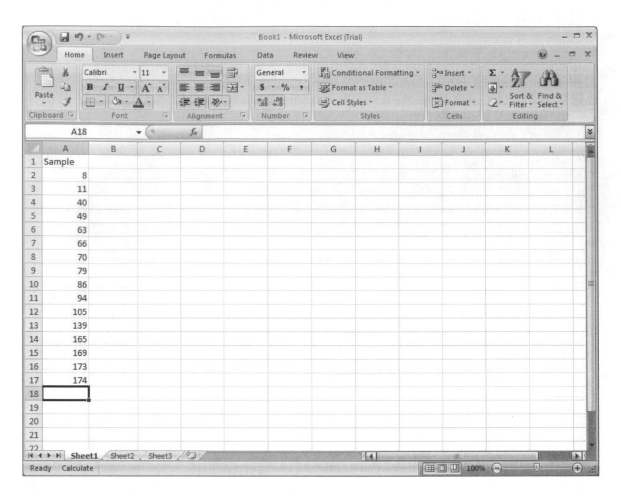

Sometimes we will want to sample from data already in our worksheet. In such a case, we can use the Sampling dialog box. To reach the Sampling dialog box, click on the **Data Analysis** button in the **Analysis** group under the **Data** tab. This will open the Data Analysis dialog box. In the **Analysis Tools** window, select **Sampling,** and click **OK.**

Example

Enter the even numbers from 0 through 200 in Column A. Then take a sample of size ten, without replacement, from the population of even numbers 0 through 200, and place the results in Column B.

First we need to enter the even numbers 0 through 200 in Column A. Let's type the label "Even #" in Cell A1. The easiest way to generate the even numbers from 0 through 200 is to use the **Fill** menu selection. To do this, we

(a) Place the value 0 into Cell A2, and finish with Cell A2 highlighted.

(b) Click the ![Fill button] (**Fill**) button, which is located on the left side of the **Editing** group under the **Home** tab. Select **Series…** from the drop-down menu. This will open the Series dialog box.

(c) In the dialog box, under **Series in,** select **Columns.** Under **Type,** select **Linear.** Enter 2 as the **Step value** and 200 as the **Stop value.** Click **OK.**

Copyright © Cengage Learning. All rights reserved.

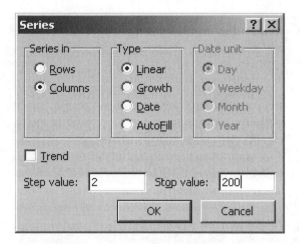

Now Column A should contain the even numbers from 0 to 200.

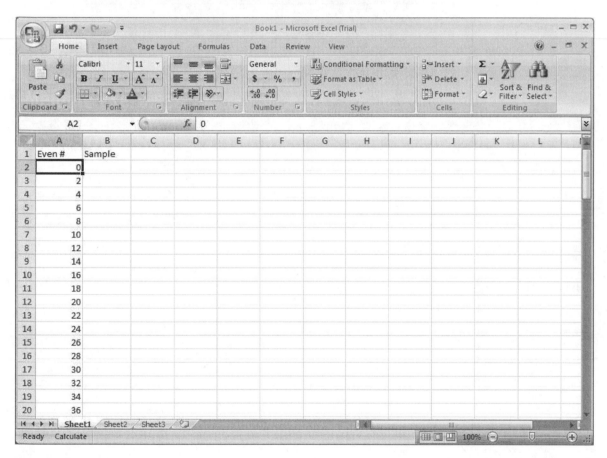

Now we will use the Sampling dialog box to select a sample of size ten from Column A, and we will place the sample in Column B. Notice that we labeled Column B as "Sample." To draw the sample,

(a) Click on the **Data Analysis** button in the **Analysis** group under the **Data** tab. Select **Sampling** in the **Analysis Tools** window in the Data Analysis dialog box. Click **OK.**

(b) In the Sampling dialog box, designate the **Input Range** from which we are sampling as $A\$2:\$A\$102$. Also, specify that the range contains a label. Select **Random** and enter 10 as the **Number of Samples.** Finally, select **Output Range** and type the destination $\$B\$2:\$B\$11$. Note that A1 and B1 already contain labels. Click **OK.**

Copyright © Cengage Learning. All rights reserved.

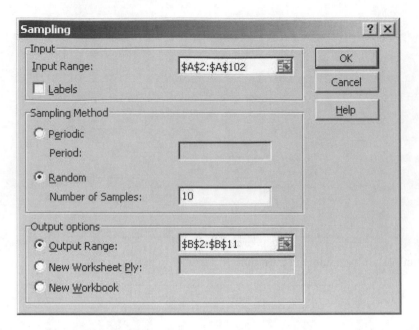

The worksheet now shows the random sample in Column B.

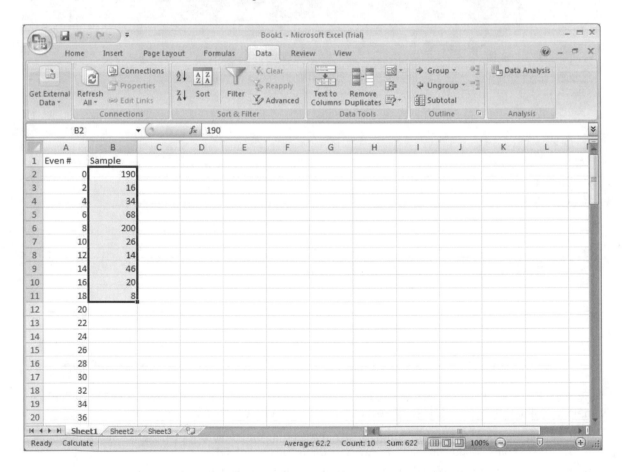

Note: After you finish the random number examples and the lab activities, you may want to set the calculation mode back to **Automatic,** especially if you are using a school computer.

Copyright © Cengage Learning. All rights reserved.

LAB ACTIVITIES FOR RANDOM SAMPLES

1. Out of a population of 8173 eligible county residents, select a random sample of fifty for prospective jury duty. (Should you sample with or without replacement?) Use the RANDBETWEEN command with bottom value 1 and top value 8173. Then sort the data to check for repetitions. Note: Be sure that the calculation mode is set to manual. Use the **Shift-F9** key combination to generate the sample in Rows 2 through 51 of Column A.

2. Retrieve the Excel worksheet Sv02.xls from the student website. This file contains weights of a random sample of linebackers on professional football teams. The data are in column form. Use the Sampling dialog box to take a random sample of ten of these weights. Print the ten weights included in the sample.

Simulating experiments in which outcomes are equally likely is another important use of random numbers.

3. We can simulate dealing bridge hands by numbering the cards in a bridge deck from 1 to 52. Then we draw a random sample of thirteen numbers without replacement from the population of 52 numbers. A bridge deck has four suits: hearts, diamonds, clubs, and spades. Each suit contains thirteen cards; those numbered 2 through 10, a jack, a queen, a king, and an ace. Decide how to assign the numbers 1 through 52 to the cards in the deck.

    (a) Use RANDBETWEEN to generate the numbers of the thirteen cards in one hand. Translate the numbers to specific cards and tell what cards are in the hand. For a second game, the cards would be collected and reshuffled. Use the computer to determine the hand you might get in a second game.

    (b) Generate the numbers 1–52 in Column A, and then use the Sampling dialog box to sample thirteen cards. Put the results in Column B, Label Column B as "My hand" and print the results. Repeat this process to determine the hand you might get in a second game.

    (c) Compare the four hands you have generated. Are they different? Would you expect this result?

4. We can also simulate the experiment of tossing a fair coin. The possible outcomes resulting from tossing a coin are heads or tails. Assign the outcome heads the number 2 and the outcome tails the number 1. Use RANDBETWEEN(1,2) to simulate the act of tossing a coin ten times. Simulate the experiment another time. Do you get the same sequence of outcomes? Do you expect to? Why or why not?

Copyright © Cengage Learning. All rights reserved.

CHAPTER 2: ORGANIZING DATA

FREQUENCY DISTRIBUTIONS AND HISTOGRAMS (SECTION 2.1 OF *UNDERSTANDING BASIC STATISTICS*)

To create a histogram in Excel, first click on the **Data Analysis** button in the **Analysis** group under the **Data** tab. This opens the Data Analysis dialog box. In the **Analysis Tools** window, select **Histogram** and click **OK.** This will open the Histogram dialog box.

Example

Let's make a histogram with four classes, using the data we stored in the workbook Book1ads (created in Chapter 1). Click on the **Office** button and select **Open** to locate the workbook, and click on the filename.

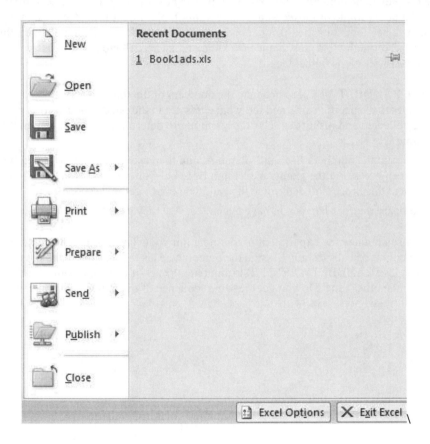

The number of ads per hour of TV is in Column A; we will represent these values in the histogram. We also need to specify class boundaries, and for this we will use Column C. Using methods shown in the text *Understanding Basic Statistics*, we see that the upper class boundaries for four classes are 16.5, 20.5, 24.5, and 28.5.

Label Column C as "Ad Count," and below the label enter these values, smallest to largest. The horizontal axis of the histogram will carry the label "Ad Count." Note: Excel follows the convention that a data value is counted in a class if the value is less than or equal to the upper boundary (upper bin value) of the class.

Open the Histogram dialog box as described above. Check **Labels** and, under the **Output options,** check **New Workbook** and **Chart Output.** Your dialog box should look similar to the one at the top of the next page.

 Copyright © Cengage Learning. All rights reserved.

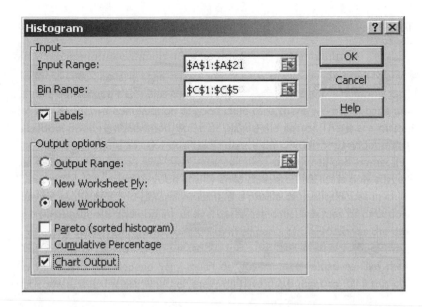

In the Input Range box, type the range of absolute addresses for Ad Count ($A$1:$A$21), and in the Bin Range box, type the range of absolute addresses for the class boundaries ($C$1:$C$5). Note that the Cells A1 and C1 are included as addresses, because the label is taken into account for the histogram analysis. Click **OK.**

In the resulting worksheet, we moved and resized the chart window. Notice that the class boundaries are not shown directly under the tick marks. The first bar represents all the values less than or equal to 16.5, the second bar is for values greater than 16.5 and less than or equal to 20.5, and so on. There are no values above 28.5.

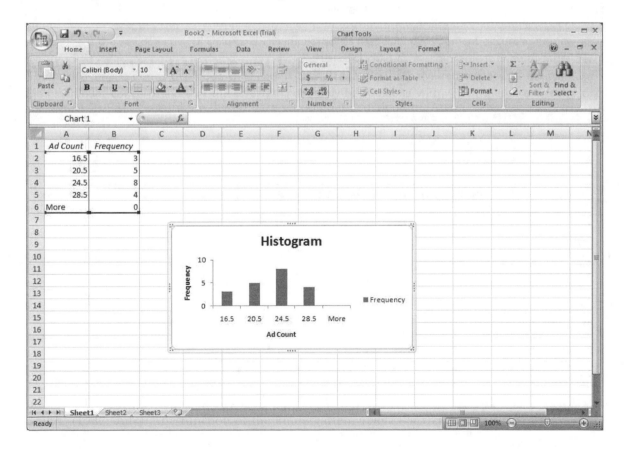

Copyright © Cengage Learning. All rights reserved.

If you do not specify the cells containing the bin range, Excel automatically creates enough bins (classes) to show the data distribution.

Adjusting the Histogram

In the histogram, Excel automatically supplies the "More" category, which will be empty if you specify upper class boundaries. To remove the "More" category, select the cells in the row of the frequency table that contains "More". Then, in the **Cells** group under the **Home** tab, select **Delete** and **Delete Sheet Rows.** To keep the size of the histogram uniform, click the **Insert** button and select **Insert Sheet Rows.**

To make the bars touch, *right-click* on one of the bars of the histogram. Then click **Format Data Series…** and under the **Series Options** option move the **Gap Width** slider to 0% (No Gap). Then click **Close.** The results should be similar to the following:

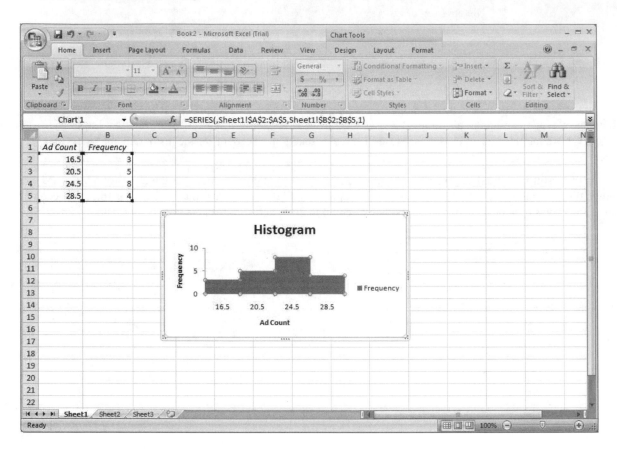

Copyright © Cengage Learning. All rights reserved.

LAB ACTIVITIES FOR FREQUENCY DISTRIBUTIONS AND HISTOGRAMS

1. The Book1ads workbook contains a second column of data, with the minutes of ads per hour of prime time TV. Retrieve the workbook again and use Column B to

   (a) Make a histogram letting Excel determine the number of classes (bins).

   (b) Use the **Sort and Filter** button in the **Editing** group under the **Home** tab to find the highest and lowest data values. Use the techniques in the text to find the upper class boundaries for five classes. Make a column in your worksheet that contains these boundaries, and label the column "Min/Hr." In the Histogram dialog box, use the column containing these upper boundaries as the bin range and generate a histogram.

2. As a project for her finance class, Linda gathered data about the number of cash requests made between the hours of 6 P.M. and 11 P.M at an automatic teller machine located in the student center. She made a count every day for four weeks. The results follow.

25	17	33	47	22	32	18	21	12	26	43	25
19	27	26	29	39	12	19	27	10	15	21	20
32	24	17	181								

   (a) Enter the data.

   (b) Repeat part (b) of Problem 1.

3. Choose one of the following workbooks from the Excel data disk.

   DISNEY STOCK VOLUME: Sv01.xls
   WEIGHTS OF PRO FOOTBALL PLAYERS: Sv02.xls
   HEIGHTS OF PRO BASKETBALL PLAYERS: Sv03.xls
   MILES PER GALLON GASOLINE CONSUMPTION: Sv04.xls
   FASTING GLUCOSE BLOOD TESTS: Sv05.xls
   NUMBER OF CHILDREN IN RURAL CANADIAN FAMILIES: Sv06.xls

   (a) Make a histogram, letting Excel scale it.

   (b) Make a histogram using five classes. Use the method in part (b) of Problem 1.

4. Histograms are not effective displays for some data. Consider the following data:

1	2	3	6	7	9	8	4	12	10
1	9	1	12	12	11	13	4	6	206

   Enter the data and make a histogram, letting Excel do the scaling. Now drop the high value, 206, from the data set. Do you get more refined information from the histogram by eliminating the high and unusual data value?

Copyright © Cengage Learning. All rights reserved.

BAR GRAPHS, CIRCLE GRAPHS, AND TIME-SERIES GRAPHS (SECTION 2.2 OF *UNDERSTANDING BASIC STATISTICS*)

Excel 2007 has a wide variety of charts that can be accessed via the **Charts** group under the **Insert** tab.

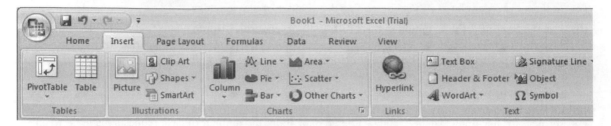

When you click on the picture of any chart type in the **Charts** group (for example, **Column**), you will see a window containing pictures of the various styles of charts available for that type.

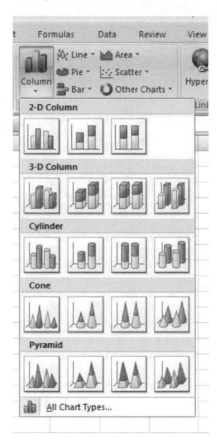

You can also view all available charts by clicking on **All Chart Types...** at the bottom of any one of the open chart windows.

Bar Graphs

You have the option of making a vertical bar graph (called a column graph in Excel) or a horizontal bar graph (called a bar graph in Excel).

Before making a chart, you must enter the necessary data in a worksheet, in rows or columns with appropriate row and column headers.

Copyright © Cengage Learning. All rights reserved.

Example

If you are out hiking, and the air temperature is 50°F with no wind, a light jacket will keep you comfortable. However, if a wind comes up, you will feel cold, even though the temperature has not changed. This is called wind chill. In the following spreadsheet, wind speeds and equivalent temperatures as a result of wind chill are given for a calm-air temperature of 50°F.

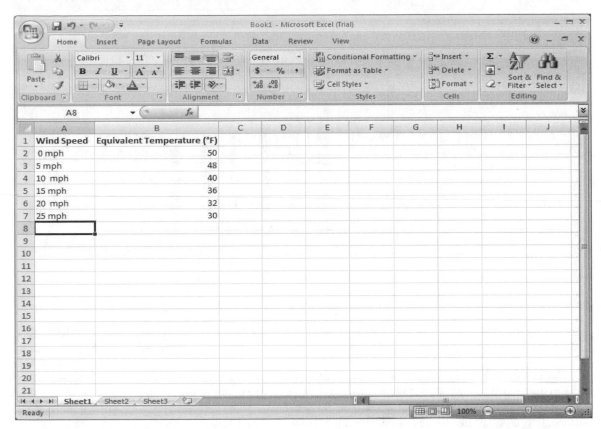

Notice that the Columns A and B are widened, and the labels are typed in bold. Also, each wind speed is followed by a label. This will cause Excel to treat the wind speeds as row headers, rather than as numerical values. Now, after making sure that a cell in or touching the data blocks is selected, click on the **Column** chart button, choose the first option in the 2-D Column row (Clustered Column), and produce a sample chart. The row headers, i.e. the wind speeds, give us the labels on the horizontal axis, while the values in Column B, the equivalent temperatures, appear as bar heights.

Copyright © Cengage Learning. All rights reserved.

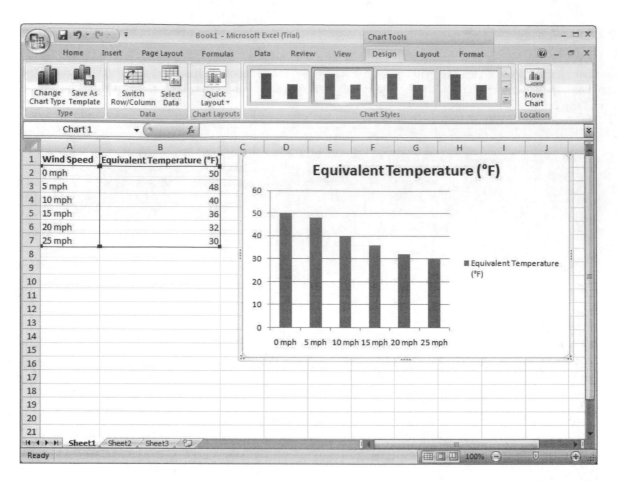

Notice that when the chart is created, extra tabs appear above the Ribbon. These tabs— **Design, Layout, and Format**—will allow you to make changes to the visual appearance of the chart. Click on these tabs to see options for labeling the axes, setting grids, labeling the heights of the bars, etc. Try different buttons and options. If you do not like the results, deselect the options. In the following image, a new title, axis labels, and column height values have been inserted in the chart.

Copyright © Cengage Learning. All rights reserved.

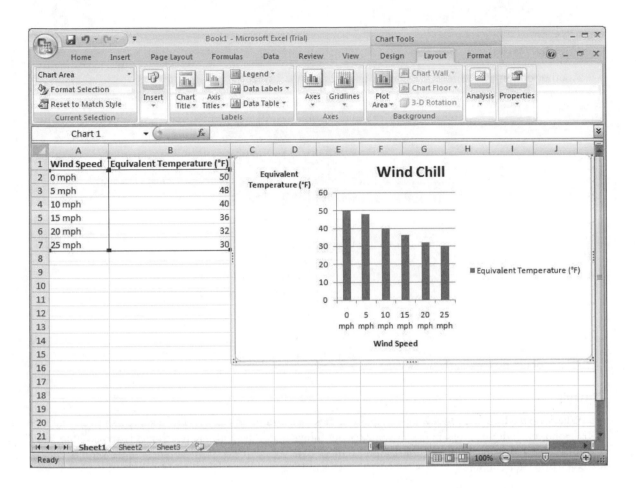

One other option worth mentioning is **Move Chart.** The **Move Chart** button may be found in the **Location** group under the **Design** tab. Clicking this button opens the Move Chart dialog box. By selecting the **New sheet** option, we will get the chart on a worksheet all by itself. Selecting the **Object in** option places the chart on the sheet designated in the window to the right.

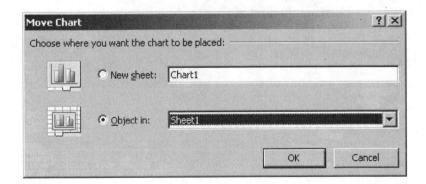

The size of the chart can be adjusted by placing the cursor on the sides or corners of the border. The chart can be moved to any part of the sheet by clicking on an empty area within the border and dragging the chart to the desired location.

Copyright © Cengage Learning. All rights reserved.

Circle Graphs

Another option available in the **Charts** group under the **Insert** tab is the Pie Chart.

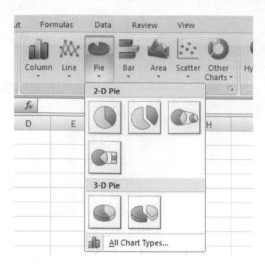

Again, enter the data in the worksheet first.

Example

Where do we hide the mess when company comes? According to *USA Today,* a survey showed that 68% of the respondents hide the mess in the closet, 23% put things under the bed, 6% put things in the bathtub, and 3% put things in the freezer. Make a circle graph for these data. We will put location labels in Column A and the percent values in Column B.

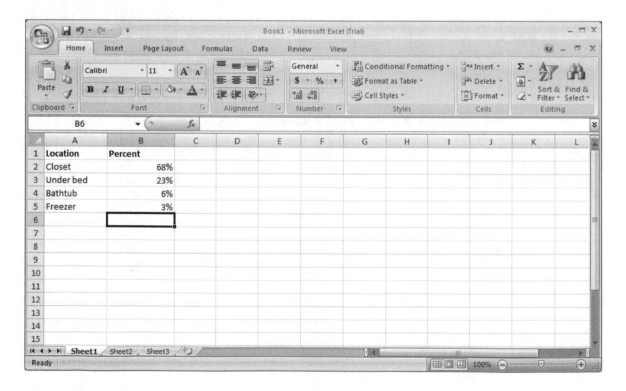

Copyright © Cengage Learning. All rights reserved.

Highlight any cell containing a data value. Then click on the **Pie** button in the **Charts** group under the **Insert** tab. Select the first option in the 2-D Pie line. A pie chart will appear on the worksheet.

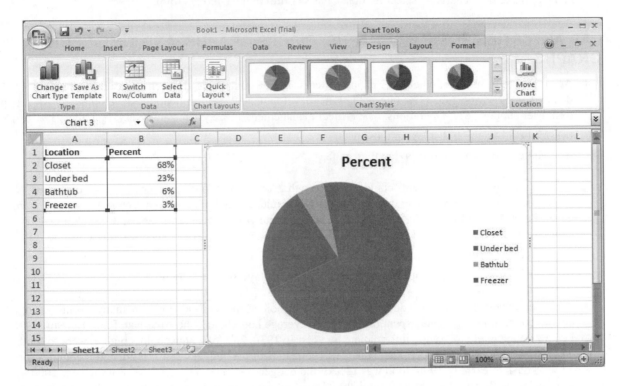

For the title, click the title box in the chart that contains the title "Percent," and replace the text with the new title type "Where We Hide the Mess." Click the **Outside End** option under the **Data Labels** button in the **Labels** group under the **Layout** tab. This will show the percentage values for each wedge of the chart.

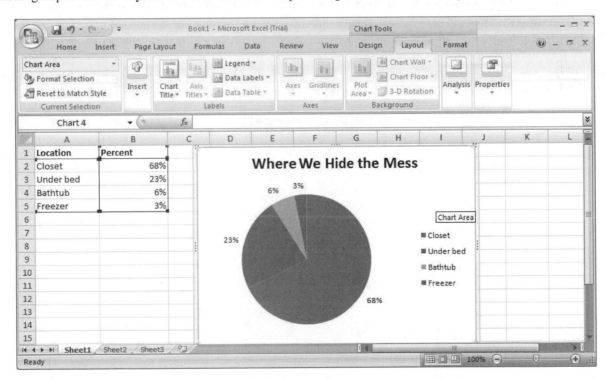

Copyright © Cengage Learning. All rights reserved.

Time-Series Graphs

You can make time charts by selecting Line Charts (**Line**) in the **Charts** group.

LAB ACTIVITIES FOR BAR GRAPHS, CIRCLE GRAPHS, AND TIME-SERIES GRAPHS

1. According to a survey of chief information officers at large companies, the technology skills most in demand are: Networking, 33%; Internet/intranet development, 21%; Applications development, 18%; Help desk/user support, 8%; Operations, 6%; Project management, 6%, Systems analysis, 5%; Other, 3%.

   (a) Make a bar graph displaying these data.

   (b) Make a circle graph displaying these data.

2. In a survey in which respondents could name more than one choice, on-line Internet users were asked where they obtained news about current events. The results are: Search engine/directory sites, 49%; Cable news site, 41%; On-line service; 40%; Broadcast news site, 40%; Local newspapers, 30%; National newspaper site; 24%; Other, 13%; National newsweekly site, 12%; Haven't accessed news on-line, 11%.

   (a) Make a horizontal bar graph displaying this information.

   (b) Is this information appropriate for a circle graph display? Why or why not?

3. What percentage of its income does the average household spend on food, and how many workdays are devoted to earning the money spent on food in an average household? The American Farm Bureau Federation gave the following information, by year: In 1930, 25% of a household's budget went to food, and it took 91 workdays to earn the money. In 1960, 17% of the budget was for food, and the money took 64 workdays to earn. In 1990, food was 12% of the budget, earned in 43 workdays. For the year 2000, it was projected that the food budget would be 11% of total income and that it would take 40 workdays to earn the money.

   (a) Enter these data in an Excel worksheet so you can create graphs.

   (b) Make bar charts for both the percent of budget for food, by year, and for the workdays required.

   (c) Make a "double" bar graph that shows side-by-side bars, by year, for the percent of budget and for the number of days of work. (You may need to change the format of the first column of numbers to something other than percent.)

   (d) Are these data suitable for a time plot? If so, use the Line graph option to create a time plot that shows both the percent of budget and the number of workdays needed to provide household food.

Copyright © Cengage Learning. All rights reserved.

CHAPTER 3: AVERAGES AND VARIATION

MEASURES OF CENTRAL TENDENCY AND VARIATION (SECTIONS 3.1 AND 3.2 OF *UNDERSTANDING BASIC STATISTICS*)

Sections 3.1 and 3.2 of *Understanding Basic Statistics* describe some of the measures used to summarize the character of a data set. Excel supports these descriptive measures.

At the left side of the Formula bar, click the **Insert Function** button:

In the Insert Function dialog box that appears, select Statistical in the **Or select a category** window, then select AVERAGE in the **Select a function** window and click **OK**. The function AVERAGE allows you to calculate the arithmetic mean of a set of numbers.

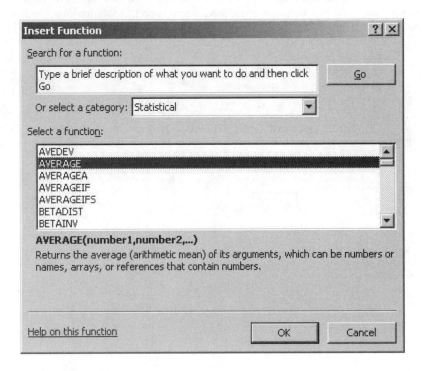

Notice that the bottom of the dialog box contains a brief explanation of how the selected function works.

Of the descriptive measures discussed in Sections 3.1 and 3.2 of the text, Excel supports the following:

AVERAGE(number1,number2,…)
  Returns the average (arithmetic mean) of its arguments, which can be numbers or names, arrays, or references that contain numbers.

COUNT(value 1,value 2,…)
  Counts the number of cells in a range that contain numbers.

MEDIAN(number1,number2,…)
  Returns the median, or the number in the middle of the set of given numbers.

MODE(number1,number2,…)
  Returns the most frequently occurring, or repetitive, value in an array or range of data.

Copyright © Cengage Learning. All rights reserved.

STDEV(number1,number2,…)
    Estimates standard deviation based on a sample (ignores logical values and text in the sample).

STDEVP(number1,number2,…)
    Calculates standard deviation based on the entire population given as arguments (ignores logical values and text).

TRIMMEAN(array,percent)
    Returns the mean of the interior portion of a set of data values.

Formatting the Worksheet to Display the Summary Statistics

It is a good idea to create a column in which you type the name of each descriptive measure you use, next to the column where the corresponding computations are performed.

Example

Let's again use the data about ads during primetime TV. We will retrieve Book1ads and find summary statistics for Column B, the time taken up with ads. We will use Column C as our column of names for the descriptive statistics measures. Notice that we widened the column to accommodate the names. We label Column C to remind us that the summary statistics apply to the number of minutes per hour that ads consume.

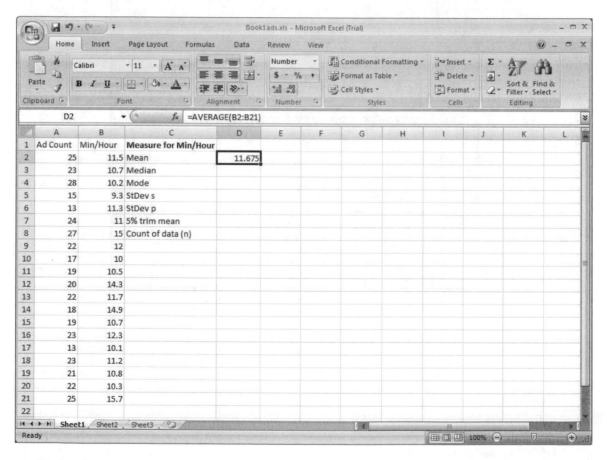

Notice that Cell D2 is highlighted, and that the Formula bar shows the command =AVERAGE(B2:B21). The value in Cell D2 is the mean of the data in the Cells B2 through B21.

Copyright © Cengage Learning. All rights reserved.

To compute the other measures, we enter the appropriate formulas in Column D and identify the measures used in Column C, as shown. Notice that you can type the commands directly in the Formula bar (don't forget to put = before the command). You also can use the **Insert Function** button, select the function from the **Select a function** window under the category Statistical, click **OK,** and insert the range of values (B2:B21) in the Function Arguments dialog box. For the 5% Trimmed Mean, the percent value is the total percentage of removed data points (10%) expressed as a decimal, or 0.1.

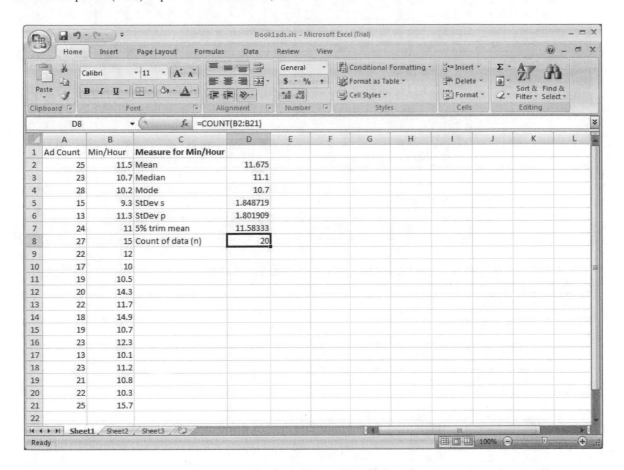

Don't forget that you can control the number of digits displayed after the decimal by using the (**Increase Decimal**) button or the (**Decrease Decimal**) button in the **Number** group under the **Home** tab. These buttons allow you to increase or decrease the number of decimal places of the value displayed in a cell.

Another way to obtain a table of some descriptive statistics is to click on the **Data Analysis** button in the **Analysis** group under the **Data** tab. This will open the Data Analysis dialog box. In the **Analysis Tools** window, select **Descriptive Statistics,** and click **OK.** This opens the Descriptive Statistics dialog box. Select $A$1:$B$21 for the **Input Range, Columns** in the **Grouped By** category, **Labels in the first row,** $D$1 for the **Output Range,** and **Summary statistics** to provide an output table containing the mean, median, mode, and other measures.

Copyright © Cengage Learning. All rights reserved.

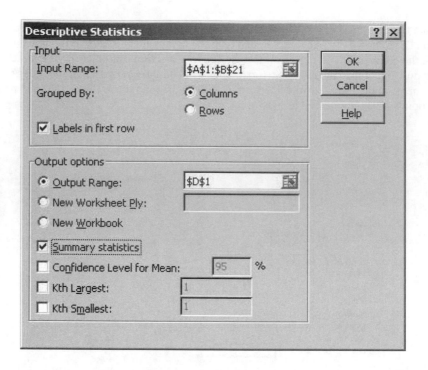

Click **OK.** For the Ad count and Min/Hour data the results are as follows:

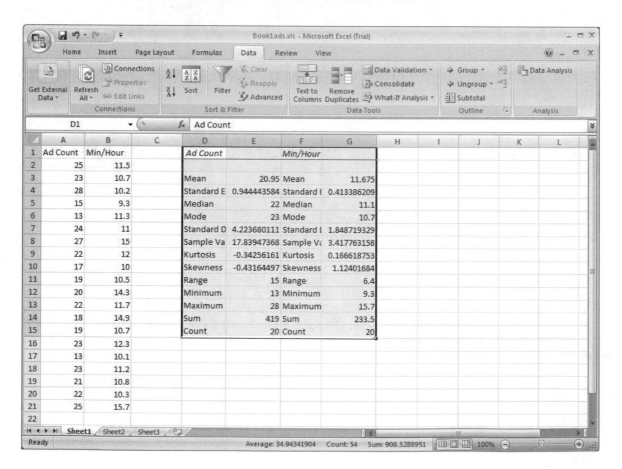

Copyright © Cengage Learning. All rights reserved.

LAB ACTIVITIES FOR CENTRAL TENDENCY AND VARIATION

1. Twenty randomly sampled people were asked to dial thirty telephone numbers each. The incidence of numbers misdialed by these people is as follows:

3	2	0	0	1	5	7	8	2	6
0	1	2	7	2	5	1	4	5	3

   Enter the data and use the appropriate commands to find the mean, median, mode, sample standard deviation, population standard deviation, 10% trimmed mean, and data count.

2. Consider the test scores of thirty students in a political science class.

85	73	43	86	73	59	73	84	100	62
75	87	70	84	97	62	76	89	90	83
70	65	77	90	94	80	68	91	67	79

   (a) Use the appropriate commands to find the mean, median, mode, sample standard deviation, 10% trimmed mean, and data count.

   (b) Suppose that Greg, a student in a political science course, missed a several classes because of illness. Suppose he took the final exam anyway and made a score of 30 instead of 85 as listed in the data set. Change the 85 (first entry in the data set) to 30 and use the appropriate commands to find the new mean, median, mode, sample standard deviation, 10% trimmed mean, and data count. Compare the new mean, median and standard deviation with the ones in part (a). Which average was most affected: median or mean? What about the standard deviation?

3. Consider the following ten data values:

4	7	3	15	9	12	10	2	9	10

   (a) Use the appropriate commands to find the sample standard deviation and the population standard deviation. Compare the two values.

   (b) Now consider these fifty data values in the same general range:

7	9	10	6	11	15	17	9	8	2
2	8	11	15	14	12	13	7	6	9
3	9	8	17	8	12	14	4	3	9
2	15	7	8	7	13	15	2	5	6
2	14	9	7	3	15	12	10	9	10

   Again, use the appropriate commands to find the sample standard deviation and the population standard deviation. Compare the two values.

   (c) Compare the results of parts (a) and (b). As the sample size increases, does it appear that the difference between the population and sample standard deviations decreases? Why would you expect this result from the formulas?

4. In this problem we will explore the effects of changing data values by multiplying each data value by a constant, or by adding the same constant to each data value.

   (a) Clear your workbook or begin a new one. Then enter the following data into Column A, with the column label "Original" in Cell A1.

1	8	3	5	7	2	10	9	4	6	3

Copyright © Cengage Learning. All rights reserved.

(b) Now label Column B as "A * 10." Select Cell B2. In the Formula bar, type

=A2*10

and press Enter. Select Cell B2 again and move the cursor to the lower right corner of Cell B2. A small black + should appear. Click-drag the + down the column so that Column B contains all the data of Column A, but with each value in Column A multiplied by 10.

(c) Now suppose we add 30 to each data value in Column A and put the new data in Column C.

First label Column C as "A + 30." Then select Cell C2. In the Formula bar, type

=A2+30

and press Enter. Then select Cell C2 again and position the cursor in the lower right corner. The cursor should change shape to a small black +. Click-drag down Column C to generate all the entries of Column A increased by 30.

(d) Predict how you think the mean and the standard deviation of Columns B and C will be similar to or different from those values for Column A. Open the Descriptive Statistics dialog box and select all three columns for the **Input Range** values, **Columns**, **Labels in the first row,** $D$1 for the **Output Range,** and **Summary statistics** to generate the mean and standard deviation values for all three columns. Compare the actual results to your predictions. What do you predict will happen to these descriptive statistics values if you multiply each data value of Column A by 50? If you add 50 to each data value in Column A?

BOX-AND-WHISKER PLOTS (SECTION 3.3 OF *UNDERSTANDING BASIC STATISTICS*)

Excel does not have any commands or dialogue boxes that produce box-and-whisker plots directly. Macros can be written to accomplish the task. However, Excel does have commands to produce the five-number summary, and you can then draw a box-and-whisker plot by hand.

The commands for the five-number summary can be found in the dialogue box obtained by pressing the **Insert Function** button on the Formula bar and selecting the following functions.

MIN(number1,number2,...)
Returns the smallest number in a set of values. Ignores logical values and text.

QUARTILE(array,1)
Returns the first quartile of a data set.

MEDIAN(number1,number2,...)
Returns the median, or the number in the middle of the set of given numbers.

QUARTILE(array,3)
Returns the third quartile of a data set.

MAX(number1,number2,...)
Returns the largest value in a set of values. Ignores logical values and text.

Copyright © Cengage Learning. All rights reserved.

Example

Generate the five-number summary for the number of minutes of ads per hour on commercial TV, using the data in Book1ads.

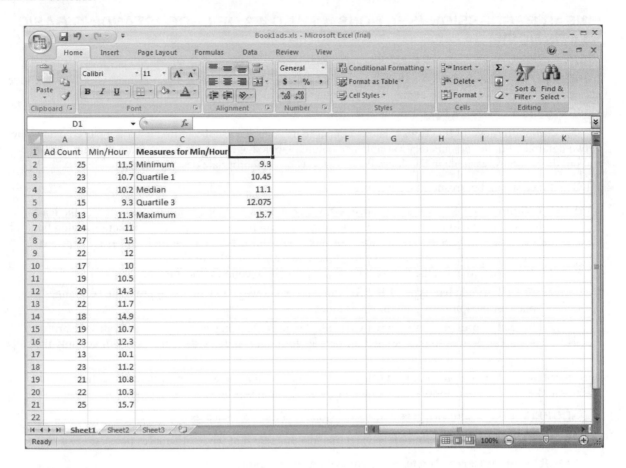

In computing quartiles, Excel uses a slightly different process from the one adopted in *Understanding Basic Statistics*. However, the results in general will be nearly the same.

Copyright © Cengage Learning. All rights reserved.

CHAPTER 4: CORRELATION AND REGRESSION

LINEAR REGRESSION (SECTIONS 4.1 AND 4.2 OF *UNDERSTANDING BASIC STATISTICS*)

Chapter 4 of *Understanding Basic Statistics* introduces linear regression. The formula for the correlation coefficient r is given in Section 4.1. Formulas to find the equation of the least-squares line, $y = a + bx$, are given in Section 4.2. This section also contains the formula for the coefficient of determination, r^2, as well as information about residuals and residual plots.

Excel supports several functions related to linear regression. To use these, first enter the paired data values in two columns. Put the explanatory variable in a column labeled with x, or an appropriate descriptive name, and put the response variable in a column labeled with y, or an appropriate descriptive name.

The functions and corresponding syntax are

LINEST(known_y's,known_x's,const,stats), which returns the slope b and y-intercept a of the least-squares line, in that order. Although this command can be found in the Insert Function dialog box, it is best to type it in because it is an array formula. (The full LINEST function involves two optional parameters, const and stats, but we will ignore these.) To use LINEST,
1. Activate two cells, the first to hold the slope b, the second for the intercept a.
2. In the Formula bar, type =LINEST(known_y'x,known_x's) with the appropriate cell ranges in place of the known x and known y values.
3. Instead of pressing Enter, press **Ctrl+Shift+Enter**. This key combination activates the array formula features so that you get the outputs for both b and a. Otherwise, you will get only the slope b of the least-squares line.

The other functions are employed in the usual way, by activating a cell and then either typing the command directly in the Formula bar, followed by Enter, or by using the Insert Function dialog box.

SLOPE(known_y's,known_x's) returns the slope b of the least-squares line.

INTERCEPT(known_y's,known_x's) returns the intercept a of the least-squares line.

CORREL(array1,array2) returns the correlation coefficient r.

FORECAST(x,known_y's,known_x's) returns the predicted y value for the specified x value, using extrapolation from the given pairs of x and y values. Note that you need to use a new FORECAST command for each different x value.

We will now go to the **Charts** group under the **Insert** tab to generate a scatter diagram and then add the results of the least-squares regression.

Example

In retailing, merchandise loss due to shoplifting, damage, and other causes is called shrinkage. The managers at H. R. Merchandise think that there is a relationship between shrinkage and the number of clerks on duty. To explore this relationship, a random sample of seven weeks was selected. During each week, the staffing level of sales clerks was held constant and the dollar value (in hundreds of dollars) of the shrinkage was recorded.

X	10	12	11	15	9	13	8	Staffing level
Y	19	15	20	9	25	12	31	(in hundreds)

Open a worksheet. Place the X values in Column A and the Y values in Column B, both with corresponding labels.

Copyright © Cengage Learning. All rights reserved.

Create a Scatter Diagram

Select the *X* and *Y* value cells (no label) for the data range. Click the **Scatter** button in the **Charts** group under the **Insert** tab. Choose the first option in the Scatter window (Scatter with only Markers), and produce a scatter diagram. Give the axes labels by selecting the **Labels** group under the **Layout** tab, which is located under the **Chart Tools** tab in the Title bar. Select **Primary Horizontal Axis Title,** choose the **Title Below Axis** option., and insert the label "X, Clerks" in the box that appears below the horizontal axis of the graph. Similarly, select **Primary Vertical Axis Title,** choose the **Rotated Title** option, and insert "Y, Shrinkage (in 100's)" in the box that appears beside the vertical axis of the graph. In the **Labels** group, click on the **Legend** button and turn off Legend by selecting **None.**

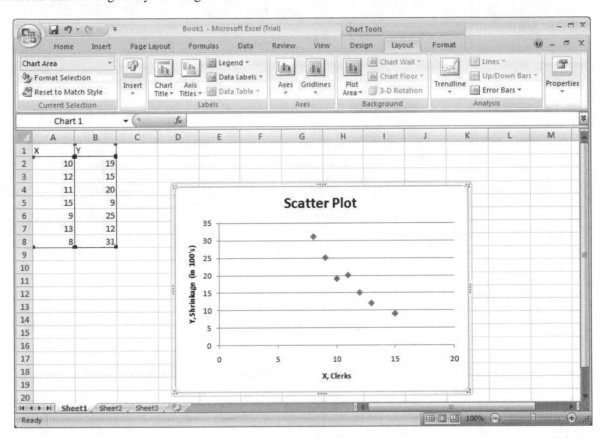

Add least-squares results to the plot

The least-squares line can be added to the plot, along with its equation and the value of r^2. Right-click on one of the data points shown in the scatter diagram. A drop-down menu will appear. Select **Add Trendline...** to call up the dialog box below. Be sure **Linear** is selected as the **Trend/RegressionType.** Check **Automatic** for the **Trendline Name,** and check **Display Equation on chart** and **Display R-squared value on chart** at the bottom of the box.

Copyright © Cengage Learning. All rights reserved.

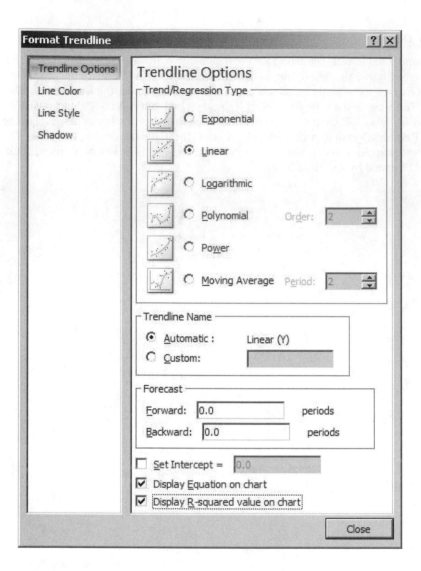

Click **Close.** Now our scatter diagram shows the graph of the least-squares line and the equation. You can move the equation out of the way of the graph by clicking on it and dragging the resulting box to a convenient location.

Copyright © Cengage Learning. All rights reserved.

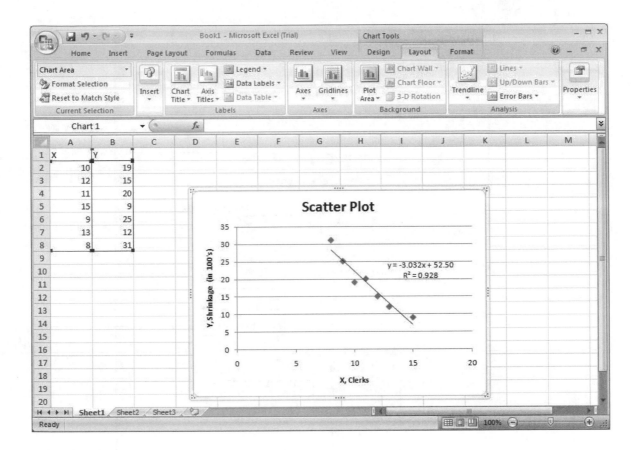

Scale the Axes

Sometimes all the points of a scatter diagram are in a corner, and we want to rescale the axes to reflect the data range. To do so,

1.   Right-click on any *X*-axis value and select **Format Axis....**

2.   In the Format Axis dialog box, change the minimum and maximum values for *X* as shown on the next page.

Copyright © Cengage Learning. All rights reserved.

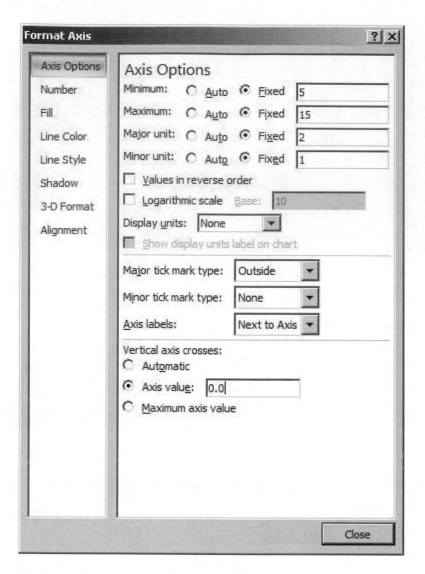

Click **Close.** A similar procedure is used to rescale the *Y*-axis. The resulting changes appear as shown at the top of the next page.

Copyright © Cengage Learning. All rights reserved.

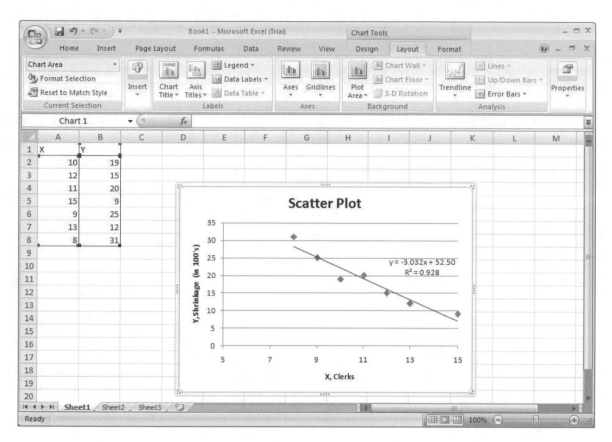

Forecast a value

Let's predict the shrinkage when fourteen clerks are available, using FORECAST. Select Cell D2 and click the **Insert Function** button on the left side of the Formula bar. Select Statistical in the **Or select a category** window, scroll down in the **Select a function** window until you reach FORECAST, and click **OK.** Fill in the Function Arguments dialog box as shown below.

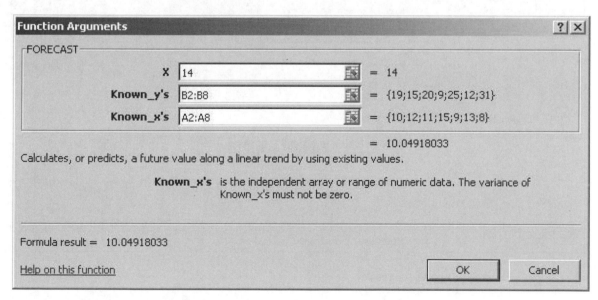

The predicted *Y* value is about 10.05. Because the unit for *Y* is hundreds of dollars, this represents a predicted shrinkage of $1005 when fourteen clerks are on duty.

Copyright © Cengage Learning. All rights reserved.

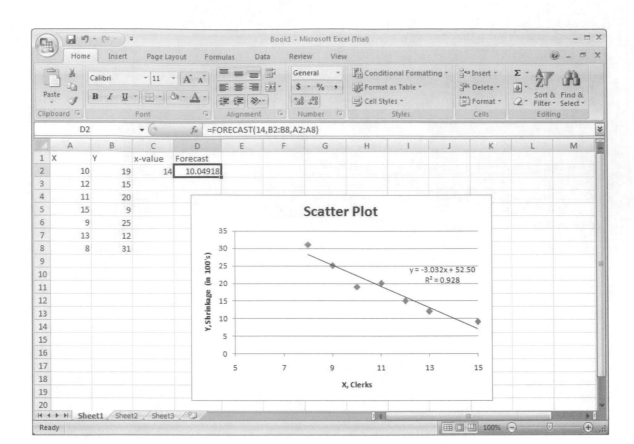

Find the value of *r*

Select a cell, click the **Insert Function** button, and under Statistical select CORREL to find the correlation coefficient.

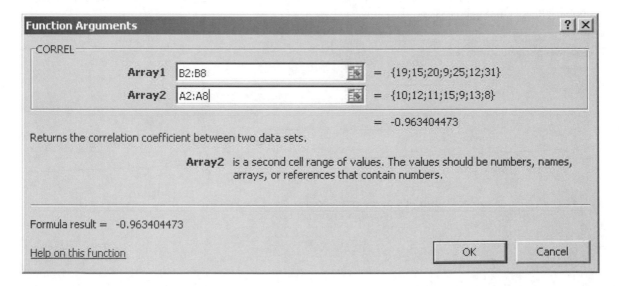

The following worksheet shows the results.

Copyright © Cengage Learning. All rights reserved.

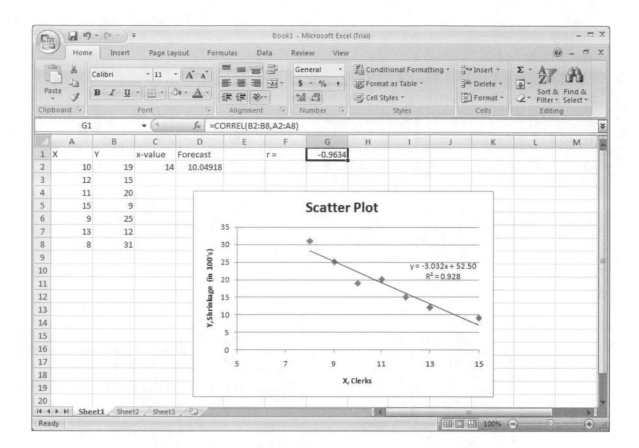

If we want to obtain the values of *b* and *a* in the least-squares line without producing a graph, we can use LINEST. Activate *two* cells on the worksheet. Type =LINEST(B2:B8,A2:A8) in the Formula bar, and press **Ctrl+Shift+Enter** to generate the values of both *b* and *a*.

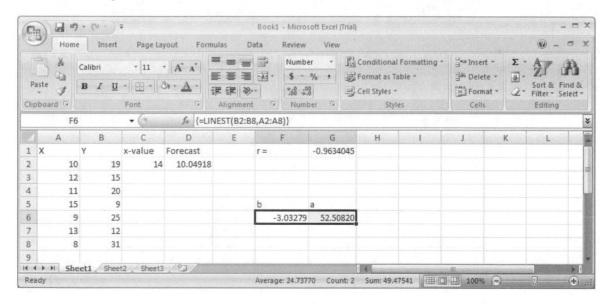

Copyright © Cengage Learning. All rights reserved.

Create a Residual Plot

A residual plot shows the amount that the predicted values for *y* deviates from the data values for *y*. Residual plots are discussed further in the Expand Your Knowledge problem 15 of Section 4.2.

A way to produce a graph showing the residuals for the various *y* values, as well as get a great deal of information at once, is with the **Regression** tool. Click the **Data Analysis** button in the **Analysis** group to open the Data Analysis dialog box, then select **Regression** in the **Analysis Tools** window. This will open the Regression dialog box. Select the options as shown below, including **Residual Plots.**

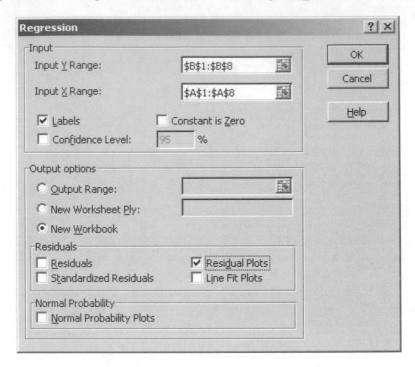

Click **OK.** Notice that the output goes to a new workbook. Select Columns A through I and click the **Format** button in the **Cells** group under the **Home** tab. Under the **Cell Size** option select **AutoFit Column Width** to adjust the column widths of the output.

Copyright © Cengage Learning. All rights reserved.

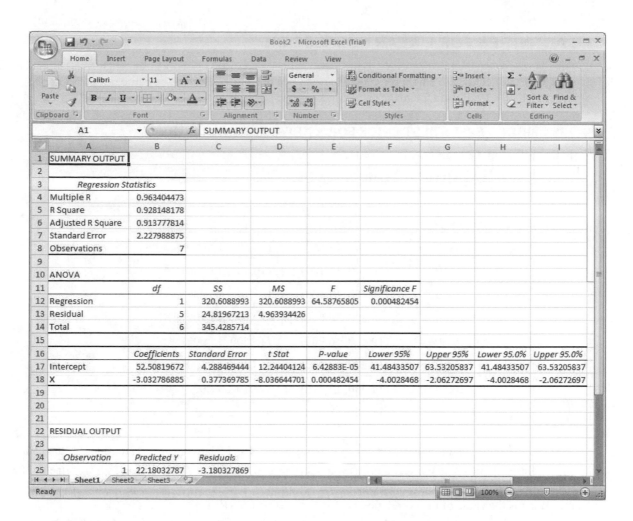

Scrolling down shows the rest of the residual output. Scrolling right shows the residual plot.

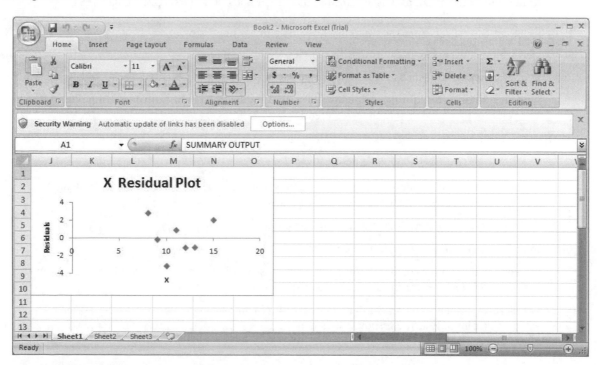

Copyright © Cengage Learning. All rights reserved.

LAB ACTIVITIES FOR LINEAR REGRESSION

1. Open or retrieve the worksheet Slr01.xls from the student website. This worksheet contains the following data, with the list price in Column C1 and the best price in Column C2. The best price is the best price negotiated by a team from the magazine.

LIST PRICE VERSUS BEST PRICE FOR A NEW GMC PICKUP TRUCK

In the following data pairs (X, Y),

X = List Price (in $1000) for a GMC Pickup Truck

Y = Best Price (in $1000) for a GMC Pickup Truck

Source: Consumers Digest, February 1994

(12.400), 11.200)	(14.300, 12.500)	(14.500, 12.700)
(14.900, 13.100)	(16.100, 14.100)	(16.900, 14.800)
(16.500, 14.400)	(15.400, 13.400)	(17.000, 14.900)
(17.900, 15.600)	(18.800, 16.400)	(20.300, 17.700)
(22.400, 19.600)	(19.400, 16.900)	(15.500, 14.000)
(16.700, (14.600)	(17.300, 15.100)	(18.400, 16.100)
(19.200, 16.800)	(17.400, 15.200)	(19.500, 17.000)
(19.700, 17.200)	(21.200, 18.600)	

(a) Use the **Scatter** button in the **Charts** group under the **Insert** tab to create a scatter plot for the data.

(b) Right-click on a data point and use the **Add Trendline...** option to show the least-squares line on the scatter diagram, along with its equation and the value of r^2.

(c) What is the value of the correlation coefficient r? (Use CORREL.)

(d) Use the least-squares model to predict the best price for a truck with a list price of $20,000. Note: Enter this value as 20, as X is assumed to be in thousands of dollars. (Use FORECAST.)

2. Other Excel worksheets appropriate to use for simple linear regression are:

Cricket Chirps versus Temperature: Slr02.xls

Source: *The Song of Insects* by Dr. G. W. Pierce, Harvard Press

> The chirps per second for the striped grouped cricket are stored in C1; the corresponding temperature in degrees Fahrenheit is stored in C2.

Diameter of Sand Granules versus Slope on Beach: Slr03.xls

Source *Physical Geography* by A.M. King, Oxford press

> The median diameter (mm) of granules of sand is stored in C1; the corresponding gradient of beach slope in degrees is stored in C2.

National Unemployment Male versus Female: Slr04.xls

Source: *Statistical Abstract of the United States*

> The national unemployment rate for adult males is stored in C1; the corresponding unemployment rate for adult females for the same period of time is stored in C2.

Select these worksheets and repeat Parts (a)–(c) of Problem 1, using Column A as the explanatory variable and Column B as the response variable.

Copyright © Cengage Learning. All rights reserved.

3. A psychologist studying the correlation between interruptions and job stress rated a group of jobs for interruption level. She selected a random sample of twelve people holding jobs from among those rated, and analyzed the people's stress level. The results follow, with X being interruption level of the job on a scale of 1 (fewest interruptions) to 20 and Y the stress level on a scale of 1 (lowest stress) to 50.

Person	1	2	3	4	5	6	7	8	9	10	11	12
X	9	15	12	18	20	9	5	3	17	12	17	6
Y	20	37	45	42	35	40	20	10	15	39	32	25

(a) Enter the X values into Column A and the Y values into Column B.

(b) Follow parts (a) through (c) of Problem 1 using the X values as the explanatory data values and the Y values as response data values.

(c) Redo Part (b). This time change the X values to the response data values and the Y values to the explanatory data values (i.e. exchange headers for the X and Y columns). How do the scatter diagrams compare? How do the least-squares equations compare? How do the correlation coefficients compare? Does it seem to make a difference which variable is the response variable and which is the explanatory variable?

4. The researcher in Problem 3 was able to add to her data. Another eleven randomly-sampled people had their jobs rated for interruption level and were then evaluated for stress level.

Person	13	14	15	16	17	18	19	20	21	22	23
X	4	15	19	13	10	9	3	11	12	15	4
Y	20	35	42	37	40	23	15	32	28	38	12

Add these data to the data in problem 3, and repeat Parts (a) and (b). Be sure to label Column A as the X values and Column B as the Y values. Compare the new correlation coefficient with the old one. Do more data tend to reduce the value of r?

Copyright © Cengage Learning. All rights reserved.

CHAPTER 5: ELEMENTARY PROBABILITY THEORY

SIMULATIONS

Excel has several random number generators. Recall from Chapter 1 that **RANDBETWEEN(bottom,top)** generates a random integer between, and including, the bottom and top numbers. Again, the Analysis ToolPak needs to be included as an Add-In to make RANDBETWEEN available. To find the RANDBETWEEN function, click the **Insert Function** button on the left side of the Formula bar. In the **Insert Function** dialog box, select All in the **Or select a category** window, scroll down in the **Select a function** window until you reach RANDBETWEEN, highlight the command, and click **OK.** You can also type the command directly in the Formula bar, but again, remember to type = first.

We can use the random number generator to simulate experiments such as tossing coins or rolling dice.

Example

Simulate the experiment of tossing a fair coin 200 times. Look at the percentage of heads and the percentage of tails. How do these compare with the expected 50% for each?

Assign the outcome heads to 1 and tails to 2. We will draw a random sample of size 200 from the distributions of integers from a minimum of 1 to a maximum of 2. When using a random number generator, you are best off setting recalculation to manual. To do this, click the **Calculation Options** button in the **Calculation** group under the **Formulas** tab, and select **Manual.**

Now put the label "Coin Toss" in Cell A1, and type =RANDBETWEEN(1,2) in Cell A2. Then press Enter. Reselect Cell A2, move the cursor to the lower right corner until the + symbol appears, and drag down through Cell A201. Because calculation is set to manual, press **Shift-F9** to apply the random integer generation command to all of the selected cells. Column A should now have 200 entries.

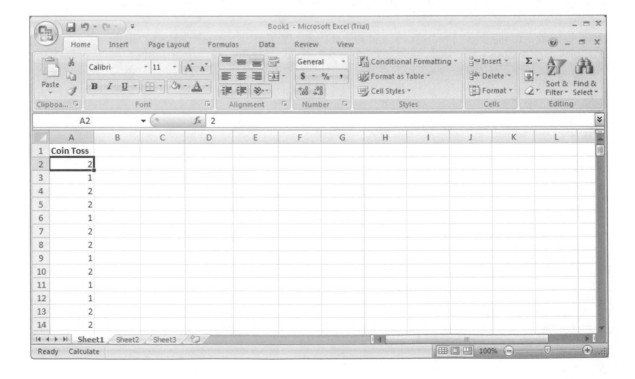

Copyright © Cengage Learning. All rights reserved.

Count the number of heads and the number of tails

Next we want to count the number of 1's and 2's in Column A. We will set up a table in Columns C and D to display the counts. Label Cell C1 as "Outcome." Type Heads in Cell C2 and Tails in Cell C3. Then Label Cell D1 as "Frequency."

We will use the COUNTIF command to count the 1's and 2's. The syntax for the COUNTIF function is

COUNTIF(range,criteria).

Recall that we assigned the number 1 to the outcome heads and the number 2 to the outcome tails. Select Cell D2, type =COUNTIF(A1:A201,1), and press Enter. This will return the number of 1's, or heads, in the designated cell range. Next select Cell D3, type = COUNTIF(A1:A201,2), and press Enter. This will return the number of 2's, or tails, in the same designated cell range.

Compute the relative frequency of each outcome

Let's use Column E to display the probability of each outcome. Label Cell E1 as "Rel. Freq." Select Cell E2. In the formula bar, type =D2/200 and press Enter. Then select Cell E3, type =D3/200, and press Enter. Notice that the relative frequencies of heads and of tails are each close to 0.5 (50%). This is what the Law of Large Numbers predicts. Of course, each time you repeat the simulation, you will obtain slightly different results.

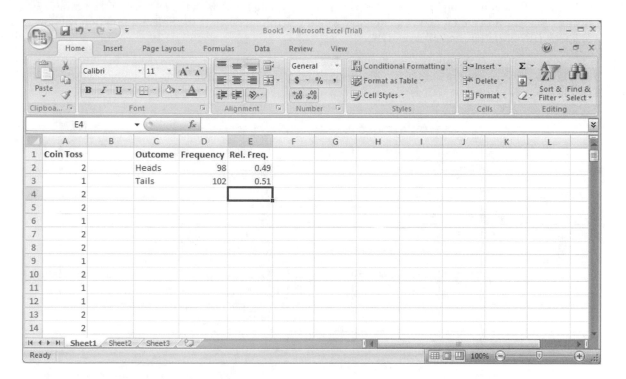

LAB ACTIVITIES FOR SIMULATIONS

1. Use RANDBETWEEN to simulate 50 tosses of a fair coin. Make a table showing the frequency of the outcomes and the relative frequency. Compare the results with the theoretical expected percentages (50% heads, 50% tails). Repeat the process for 500 trials. Are these outcomes closer to the results predicted by theory?

2. Use RANDBETWEEN to simulate 50 rolls of a fair die. Use the number 1 for the bottom value and 6 for the top. Make a table showing the frequency of each outcome and the relative frequency. Compare the results with the theoretical expected percentages (16.7% for each outcome). Repeat the process for 500 tosses. Are these outcomes closer to the results predicted by theory?

Copyright © Cengage Learning. All rights reserved.

CHAPTER 6: THE BINOMIAL PROBABILITY DISTRIBUTION AND RELATED TOPICS

THE BINOMIAL PROBABILITY DISTRIBUTION (SECTIONS 6.2 AND 6.3 OF *UNDERSTANDING BASIC STATISTICS*)

The binomial probability distribution is discussed in Chapter 6 of *Understanding Basic Statistics.* It is a discrete probability distribution controlled by the number of trials, *n,* and the probability of success on a single trial, *p.*

The Excel function that generates binomial probabilities is

BINOMDIST(number_s,trials,probability_s,cumulative),

where *r* represents the number of successes. Using TRUE for the variable cumulative returns the cumulative probability of obtaining no more than *r* successes in *n* trials. Using FALSE returns the probability of obtaining exactly *r* successes in *n* trials.

You can type the command directly in the Formula bar (don't forget the preceding = sign), or you can call up the Insert Function dialog box by clicking the **Insert Function** button on the left side of the Formula bar, selecting Statistical in the **Or select a category** window, scrolling down in the **Select a function** window until you reach BINOMDIST, and clicking **OK.**

To compute the probability of exactly three successes out of four trials where the probability of success on a single trial is 0.50, enter the following information into the Function Arguments dialog box. When you click **OK,** the formula result 0.25 will appear in the active cell.

Example

A surgeon regularly performs a certain difficult operation. The probability of success for any one such operation is $p = 0.73$. Ten similar operations are scheduled. Find the probability of success for 0 through 10 successes out of these operations.

First let's put information regarding the number of trials and probability of success on a single trial into the worksheet. We type $n = 10$ in Cell A1 and $p = 0.73$ in Cell B1.

   Copyright © Cengage Learning. All rights reserved.

Next, we will put the possible values for the number of successes, *r*, in Cells A3 through A13 with the label "r" in Cell A2. We can also duplicate A2 through A13 in D2 through D13 for easier reading of the finished table. Now place the label "P(r)" in Cell B2 and the label "P(X ≤ r)" in Cell C2. We will use Excel to generate the probabilities of the individual number of successes *r* in Cells B3 through B13 and the corresponding cumulative probabilities in Cells C3 through C13.

Generate *P*(*r*) values and adjust format

Select Cell B3 as the active cell. In the Formula bar type =BINOMDIST(A3,10,0.73,FALSE) and press Enter. Then select Cell B3 again, and move the cursor to the lower right corner of the cell. When the small black + appears, hold down the left mouse button and drag through Cell B13. Release the mouse button. This process generates the probabilities for each value of *r* in Cells A3 through A13.

The probabilities are expressed in scientific notation, where the value after the E indicates that the decimal value is multiplied by the given power of 10. To reformat the probabilities, select them all and press the comma button in the **Number** group under the **Home** tab. Then press the **Increase Decimal** button to move the decimal point until you see four digits after it.

Generate Cumulative Probabilities *P*(*X* ≤ *r*) and adjust format

Select Cell C3 as the active cell. In the Formula bar type =BINOMDIST(A3,10,0.73,TRUE) and press Enter. Then select Cell C3 again, move the cursor to the lower right corner of the cell, and drag down through Cell C13. This generates the cumulative probabilities for each value of *r* in Cells A3 through A13. Again, reformat the probabilities to show four decimal places.

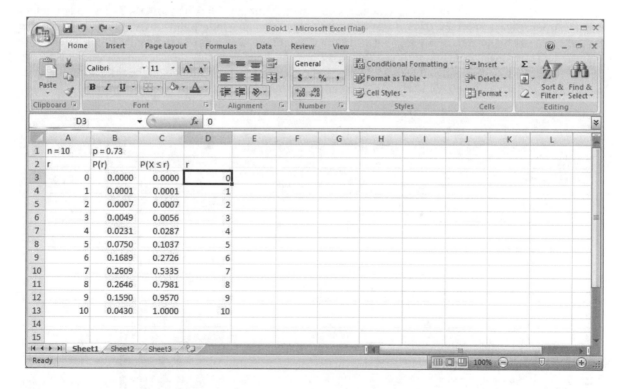

Next, use the Ribbon's **Insert** tab to create a column graph showing the probability distribution. For the data range, activate Cells B2 through B13. Press the **Column** button in the **Charts** group under the **Insert** tab. Select the first 2-D Column icon.

Copyright © Cengage Learning. All rights reserved.

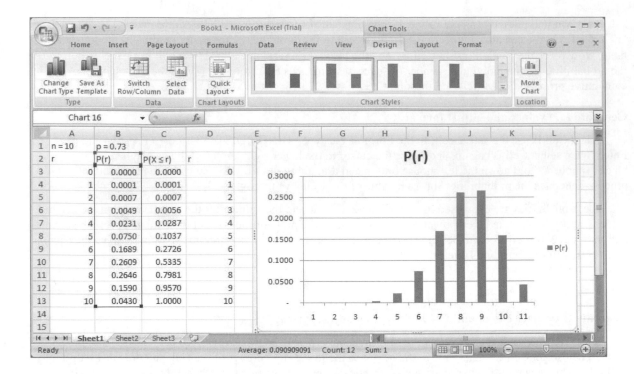

Move the graph so it doesn't block the data columns, and size it as you wish.

Under the **Chart Tools** tab in the Title bar, select the **Layout** tab. Click on the **Axis Titles** button in the **Labels** group. Select **Primary Horizontal Axis Title** and choose the **Title Below Axis** option. Type *r* in the box that appears below the horizontal axis of the graph. Similarly, select **Primary Vertical Axis Title** and choose the **Horizontal Title** option. Type $P(r)$ in the box that appears beside the vertical axis of the graph. Finally, because "P(r)" already appears in the place of the chart title, simply click on the title box and change it to "$P(r)$ for $n = 10$ and $p = 0.73$." The graph on your worksheet will appear similar to the one below.

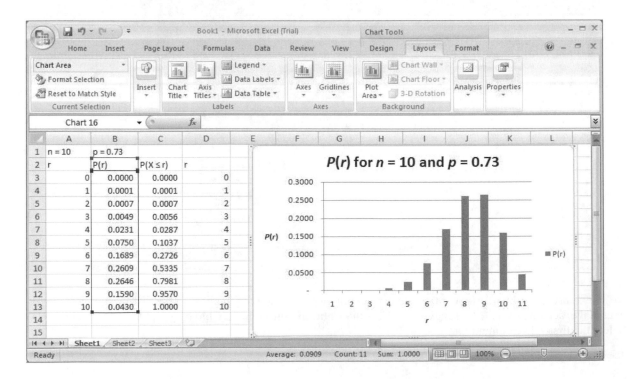

Copyright © Cengage Learning. All rights reserved.

LAB ACTIVITIES FOR BINOMIAL PROBABILITY DISTRIBUTIONS

1. You toss a coin n times. Call heads success. If the coin is fair, the probability of success p is 0.5. Use the BINOMDIST function with cumulative = FALSE to find each of the following probabilities.

   (a) Find the probability of getting exactly five heads out of eight tosses.

   (b) Find the probability of getting exactly twenty heads out of 100 tosses.

   (c) Find the probability of getting exactly forty heads out of 100 tosses.

2. You toss a coin n times. Call heads success. If the coin is fair, the probability of success p is 0.5. Use the BINOMDIST function with cumulative = TRUE to find each of the following probabilities

   (a) Find the probability of getting at least five heads out of eight tosses.

   (b) Find the probability of getting at least twenty heads out of 100 tosses.

   (c) Find the probability of getting at least forty heads out of 100 tosses.

   Hint: Keep in mind how BINOMDIST works. Which should be larger, the value in (b) or the value in (c)?

3. A bank examiner's record shows that the probability of an error in a statement for a checking account at Trust Us Bank is 0.03. The bank statements are sent monthly. What is the probability that exactly two of the next twelve monthly statements for our account will be in error? Now use the BINOMDIST function with TRUE for cumulative to find the probability that *at least* two of the next twelve statements contain errors. Use this result with subtraction to find the probability that *more than* two of the next twelve statements contain errors. You can activate a cell and use the Formula bar to do the required subtraction.

4. Some tables for the binomial distribution give values only up to 0.5 for the probability of success p. There is a symmetry between values of p greater than 0.5 and values of p less than 0.5.

   (a) Consider the binomial distribution with $n = 10$ and $p = 0.75$. Because there are anywhere from 0 to 10 successes possible, put the numbers 0 through 10 in Cells A2 through A12. Use Cell A1 for the label "r." Use BINOMDIST with cumulative = FALSE to generate the probabilities for $r = 0$ through 10. Store the results in Cells B2 through B12. Use Cell B1 for the label "p = 0.75."

   (b) Now consider the binomial distribution with $n = 10$ and $p = 0.25$. Use BINOMDIST with cumulative = FALSE to generate the probabilities for $r = 0$ through 10. Store the results in Cells C2 through C12. Use Cell C1 for the label "p = 0.25."

   (c) Now compare the entries in Columns B and C. How does $P(r = 4$ successes with $p = 0.75)$ compare to $P(r = 6$ successes with $p = 0.25)$?

5. (a) Consider a binomial distribution with fifteen trials and a probability of success on a single trial of 0.25. Create a worksheet showing values of r and their corresponding binomial probabilities. Generate a bar graph of the distribution.

   (b) Consider a binomial distribution with fifteen trials and probability of success on a single trial $p = 0.75$. Create a worksheet showing values of r and the corresponding binomial probabilities. Generate a bar graph of the distribution.

   (c) Compare the graphs of parts (a) and (b). How are they skewed? Is one symmetric with the other?

Copyright © Cengage Learning. All rights reserved.

CHAPTER 7: NORMAL CURVES AND SAMPLING DISTRIBUTIONS

GRAPHS OF NORMAL PROBABILITY DISTRIBUTIONS (SECTION 7.1 OF *UNDERSTANDING BASIC STATISTICS*)

A normal distribution is a continuous probability distribution governed by the parameters μ (the mean) and σ (the standard deviation), as discussed in Section 7.1 of *Understanding Basic Statistics*. The Excel function that generates values for a normal distribution is

$$NORMDIST(x,mean,standard\_dev,cumulative).$$

By setting the value of cumulative to FALSE, the function gives the values of the normal probability density function for the corresponding x value. You can type the NORMDIST command directly into the Formula bar (don't forget the preceding = sign), or you can call up the Insert Function dialog box by clicking the **Insert Function** button on the left side of the Formula bar, selecting Statistical in the **Or select a category** window, scrolling down in the **Select a function** window until you reach NORMDIST, and clicking **OK.** The resulting Function Arguments dialog box appears as shown below; in this case, the entries are those used in the next example.

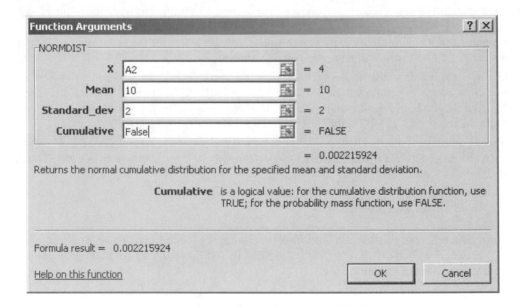

Example

Graph the normal distribution with mean $\mu = 10$ and standard deviation $\sigma = 2$.

Because most of the normal curve ranges from $\mu - 3\sigma$ to $\mu + 3\sigma$, we will start the graph at $10 - 3(2) = 4$ and end it at $10 + 3(2) = 16$. We will let Excel set the scale on the vertical axis automatically.

Generate the column of x values

To graph a normal distribution, we must have two columns containing x values and corresponding y values. We begin by generating x values from 4 to 16, with an increment of 0.25. Select Cell A2 and enter the number

4. Select Cell A2 again, click on the (**Fill**) button in the **Editing** group under the **Home** tab, and select **Series...** from the pull-down menu. In the Series dialogue box, set the options as shown on the next page.

                                                      Copyright © Cengage Learning. All rights reserved.

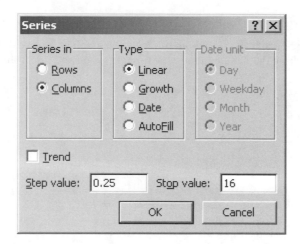

. Click **OK.** Note that Column A now contains the numbers 4, 4.25, 4.50, … all the way up to 16. Type **x** for the column heading in Cell A1.

Generate the column of *y* values

Select Cell B2, type =NORMDIST(A2,10,2,FALSE) in the Formula bar, and press Enter. Select Cell B2 again and move the cursor to the lower right corner of the cell. When the cursor changes to a small black +, hold down the left mouse button and drag down the column until each Column-A entry has a corresponding Column-B entry. Release the mouse button. All the *y* values should now appear. Use the **Decrease Decimal** button in the **Number** group to set the number of decimal places to six. Type **y** for the column heading in Cell B1.

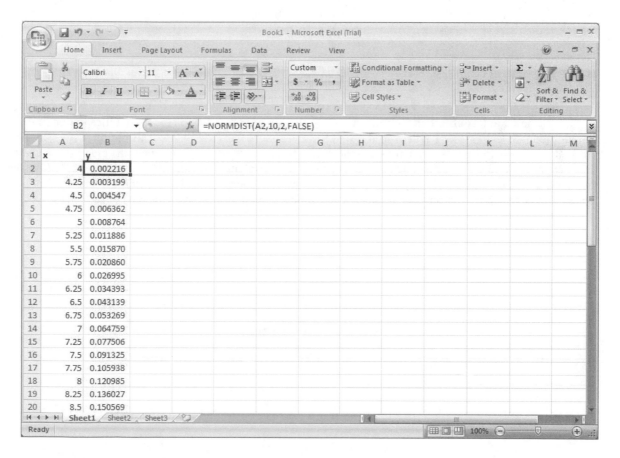

Copyright © Cengage Learning. All rights reserved.

Create the graph of a normal distribution

We will use the features under the **Insert** tab to create the graph of a normal distribution. Select Column B as the data range, then click on the **Line** button in the **Charts** group and select the first Line graph icon.

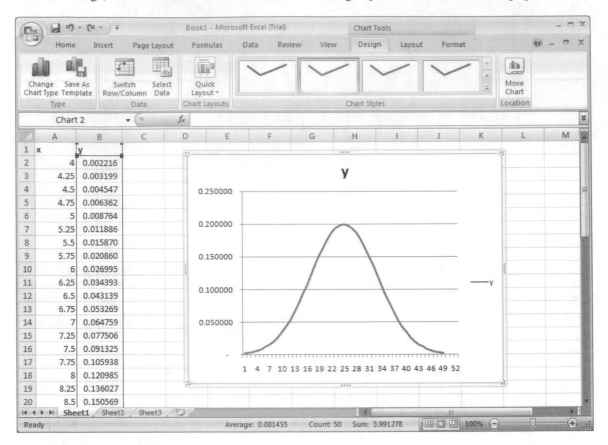

Under the **Chart Tools** tab in the Title bar, select the **Layout** tab. Click on the **Axis Titles** button in the **Labels** group. Select **Primary Horizontal Axis Title** and choose the **Title Below Axis** option. Type x in the box that appears below the horizontal axis of the graph. Because "y" already appears in the place of the chart title, simply click on the title box and change it to "Normal Distribution, Mean = 10, Std. Dev. = 2." Move the graph and adjust its size to your liking. Notice that as you make the graph wider or taller, the labels on the x-axis may change. The graph on your worksheet will appear similar to the one on the next page.

Copyright © Cengage Learning. All rights reserved.

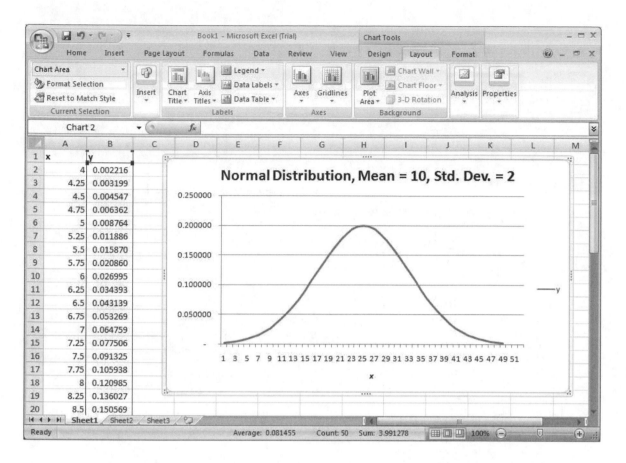

We can also graph two (or more) normal distributions on the same graph. In the next display, we generated *y* values for a normal distribution with mean 10 and standard deviation 1 in Column C. Select Columns B and C for the data range. Then click the **Line** button in the **Charts** group and select the first Line graph icon.

Copyright © Cengage Learning. All rights reserved.

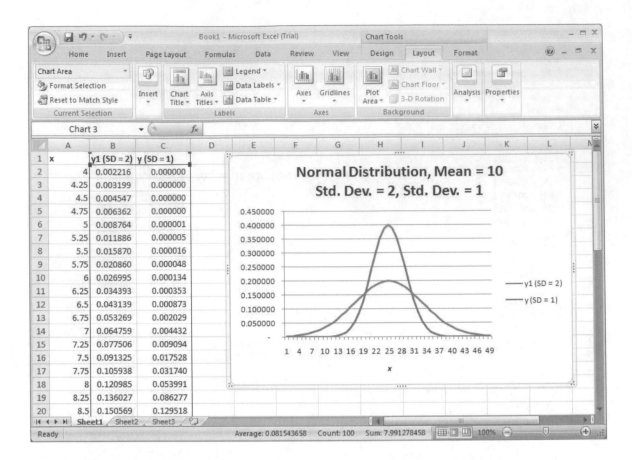

STANDARD SCORES AND NORMAL PROBABILITIES (SECTION 7.2 OF *UNDERSTANDING BASIC STATISTICS*)

Excel has several built-in functions relating to normal distributions.

STANDARDIZE(x,mean,standard_dev) returns the z score for the given x value from a distribution with the specified mean and standard deviation.

NORMDIST(x,mean,standard_dev,cumulative) When cumulative is TRUE, this returns probability that a random value x selected from this distribution is $\leq x_1$, i.e. it returns $P(x \leq x_1)$. This is the same as the area to the left of the specified x_1 value under the described normal distribution. When cumulative is FALSE, it returns the height of the normal probability density function evaluated at x_1. We used this function to graph a normal distribution.

NORMINV(probability,mean,standard_dev) returns the inverse of the normal cumulative distribution. In other words, when a probability is entered, the command returns the value x from the normal distribution with specified mean and standard deviation so that the area to the left of x is equal to the designated probability.

NORMSDIST(z) returns the probability that a randomly selected z score is less than or equal to the specified value of z_1, i.e. it returns $P(z \leq z_1)$. This is the same as the area to the left of the specified z_1 value under the standard normal distribution. This command is equivalent to NORMDIST(x,0,1,TRUE).

NORMSINV(probability) returns the value z such that the area to its left under the standard normal distribution is equal to the specified probability. This command is equivalent to NORMINV(probability,0,1).

Each of these commands can be typed directly into the Formula bar for an active cell or accessed by using the **Insert Function** button.

Copyright © Cengage Learning. All rights reserved.

Examples

(a) Consider a normal distribution with mean 100 and standard deviation 15. Find the z score corresponding to $x = 90$ and find the area to the left of 90 under the distribution.

First, place some headers and labels on the worksheet. Enter the value 90 in Cell A3, then

1. in Cell B3, type =STANDARDIZE(90,100,15) and press Enter.

2. in Cell C3, type =NORMDIST(90,100,15,TRUE) and press Enter.

Adjust the number of decimal places to five.

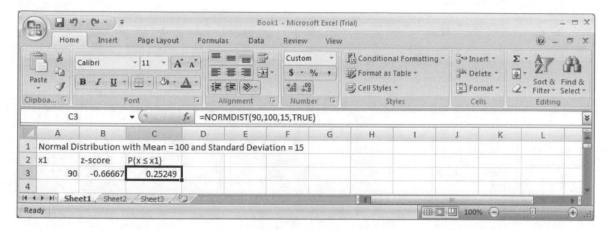

(b) Find the z score so that 10% of the area under the standard normal distribution is to the left of z.

Again, put some labels on the worksheet and enter the value 0.1 in Cell A3, Because this example involves a standard normal distribution, use =NORMSINV(0.1) and press Enter. Again, adjust the number of decimal places to five.

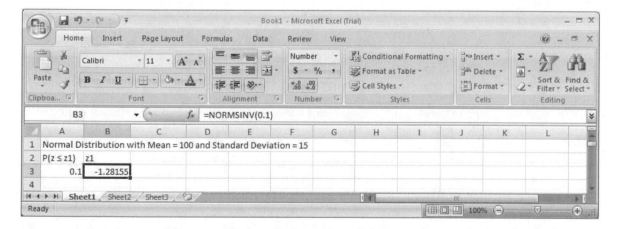

To find areas under normal curves between two values, we do simple arithmetic with the cumulative areas provided by Excel. For instance, to find the area under a standard normal distribution between –2 and 3, we would use NORMSDIST(3) to find the cumulative area to the left of 3 and NORMSDIST(–2) to find the cumulative area to the left of –2. Then we would subtract the result NORMSDIST(–2) from NORMSDIST(3). In the worksheet shown at the top of the next page, the number of decimal places has been adjusted to five.

Copyright © Cengage Learning. All rights reserved.

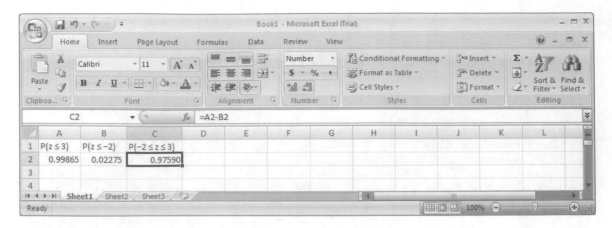

To find areas under normal curves to the right of a specified value, we use NORMDIST to find the cumulative area to the left of the value and subtract this quantity from 1. For instance, consider the normal distribution with mean 50 and standard deviation 5. Below is a worksheet in which the area to the right of 60 is found.

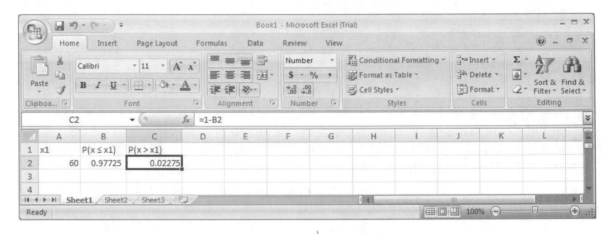

\

LAB ACTIVITIES FOR NORMAL DISTRIBUTIONS

1. **(a)** Create a graph of the standard normal distribution with a mean of 0 and a standard deviation of 1. Generate x values in Column A ranging from -3 to 3 in increments of 0.5. Use NORMDIST to generate the y values in Column B.

    (b) Create a graph of a normal distribution with a mean of 10 and a standard deviation of 1. Generate x values in Column A ranging from 7 to 13 in increments of 0.5. Use NORMDIST to generate the y values in Column B. Compare the graphs of parts (a) and (b). Do the height and spread of the graphs appear to be the same? What is different? Why would you expect this difference.

    (c) Create a graph of a normal distribution with a mean of 0 and a standard deviation of 2. Generate x values in Column A ranging from -6 to 6 in increments of 0.5. Use NORMDIST to generate the y values in Column B. Compare that graph to that of part (a). Do the height and spread of the graphs appear to be the same? What is different? Why would you expect this difference? Note: to really compare the graphs, it is best to graph them using the same scales. Redo the graph of part (a) using x from -6 to 6. Then redo the graph in this part using the same x values as in part (a) and y values ranging from 0 to the high value in part (a).

Copyright © Cengage Learning. All rights reserved.

2. Use NORMDIST or NORMSDIST plus arithmetic to find the specified area.

    (a) Find the area to the left of 2 on a standard normal distribution.

    (b) Find the area to the left of -1 on a standard normal distribution.

    (c) Find the area between -1 and 2 on a standard normal distribution.

    (d) Find the area to the right of 2 on a standard normal distribution.

    (e) Find the area to the left of 40 on a normal distribution with $\mu = 50$ and $\sigma = 8$.

    (f) Find the area to the left of 55 on a normal distribution with $\mu = 50$ and $\sigma = 8$.

    (g) Find the area between 40 and 55 on a normal distribution with $\mu = 50$ and $\sigma = 8$.

    (h) Find the area to the right of 55 on a normal distribution with $\mu = 50$ and $\sigma = 8$.

3. Use NORMINV or NORMSINV to find the specified x or z value.

    (a) Find the z value so that 5% of the area under the standard normal curve falls to the left of z.

    (b) Find the z value so that 15% of the area under the standard normal curve falls to the left of z.

    (c) Consider a normal distribution with a mean of 10 and a standard deviation of 2. Find the x value so that 5% of the area under the normal curve falls to the left of x.

    (d) Consider a normal distribution with a mean of 10 and a standard deviation of 2. Find the x value so that 15% of the area under the normal curve falls to the left of x.

SAMPLING DISTRIBUTIONS AND THE CENTRAL LIMIT THEOREM (SECTIONS 7.4 AND 7.5 OF *UNDERSTANDING BASIC STATISTICS*)

A sampling distribution is a probability distribution of a sample test statistic based on all possible simple random samples of the same size from the same population, as discussed in Section 7.4 of *Understanding Basic Statistics*. Regardless of the distribution of a variable x, the sample mean $\bar{x}$ based on a random sample of size n will have a distribution that approaches normal, as discussed in Section 7.5.

Example

Let us draw 100 random samples of size 40 from the uniform distribution on the interval from 0 to 9. We put the data into 40 columns. Then we take the mean of each of the 100 rows (40 columns across) and store them in another column. Then we can study the data of sample mean $\bar{x}$. We expect to see that the mean of the data of $\bar{x}$ is close to 4.5 and the standard deviation is close to $\sigma / n = 2.598 / \sqrt{40} = 0.411$.

In Excel 2007, click the **Data Analysis** button in the **Analysis** group under the **Data** tab. This will open the Data Analysis dialog box. In the **Analysis Tools** window, select **Random Number Generation**.

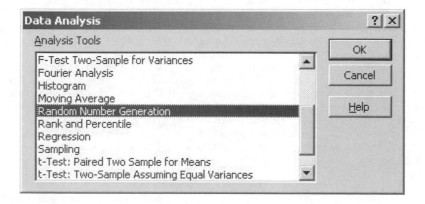

Click **OK.** This in turn opens the Random Number Generation dialog box. Enter the values shown in the box shown on the next page.

Copyright © Cengage Learning. All rights reserved.

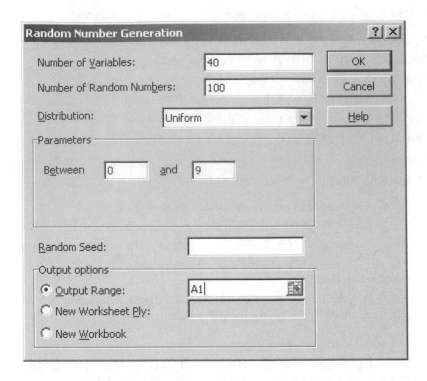

Click **OK.** A table similar to the following will appear. Note that the last column is AN and the last row is 100. The values have also been adjusted for six decimal places.

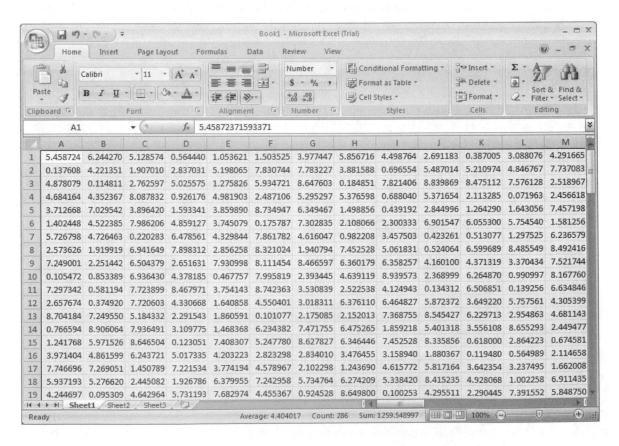

Copyright © Cengage Learning. All rights reserved.

To prepare for the next step, activate Cell A1, click the **Insert** button in the **Cells** group under the **Home** tab, and select the **Insert Sheet Rows** option. This will insert an empty row of cells and move the first line of data to Row 2. Then in Cells AP1, AQ1, and AR1, type the label "sample mean," "mean of AP," and "Std. Dev. of AP," respectively. Adjust the width of the cells accordingly.

Now in the Column AP, calculate the average of the 40 columns, which gives the data of 100 sample means. This can be done by activating Cell AP2, typing the function =AVERAGE(A2:AN2) in the Formula bar, and pressing Enter. This will produce the mean for the first row of values. To generate the means for all 100 rows, select Cell AP2 again, move the cursor to the lower right corner of the cell, and drag down through Cell AP101. This will provide the mean values from which to calculate the mean and standard deviation of these data of $\bar{x}$. Finally, select Cell AQ2 and use =AVERAGE(AP2:AP101) to find the mean of the mean values in Column AP. Then select Cell AR2 and use =STDEV(AP2:AP101) to calculate the standard deviation for the values in Column AP. As you can see, for these data the mean = 4.510241, which is close to 4.5, and the standard deviation = 0.380092, which is close to $\sigma/n = 0.411$.

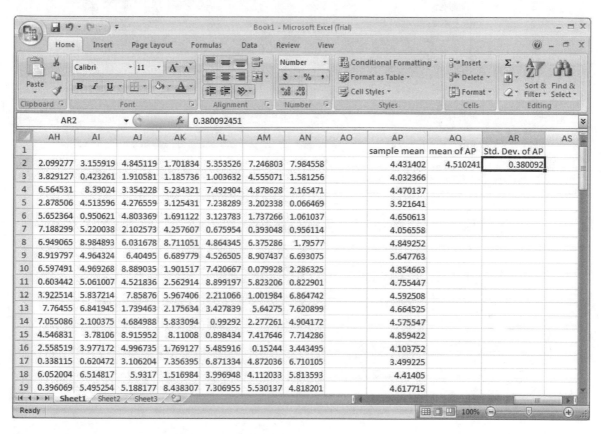

Because sample size $n = 40$, the sample mean $\bar{x}$ should approximately have a normal distribution. First, select intervals 3.5, 4, 4.5, 5, and 5.5, and enter these in Cells AQ6 through AQ10. Type the label "bound" (for boundaries) in Cell AQ5. Now click the **Data Analysis** button in the **Analysis** group under the **Data** tab to open the Data Analysis dialog box. Select **Histogram** from the **Analysis Tools** list and click **OK** to open the Histogram dialog box. Then select and type $AP$1:$AP$101 for the **Input Range**, $AQ$5:$AQ$10 for the **Bin Range,** and $AR$10 for the **Output Range.** Check the **Labels** and **Chart Output** options.

Copyright © Cengage Learning. All rights reserved.

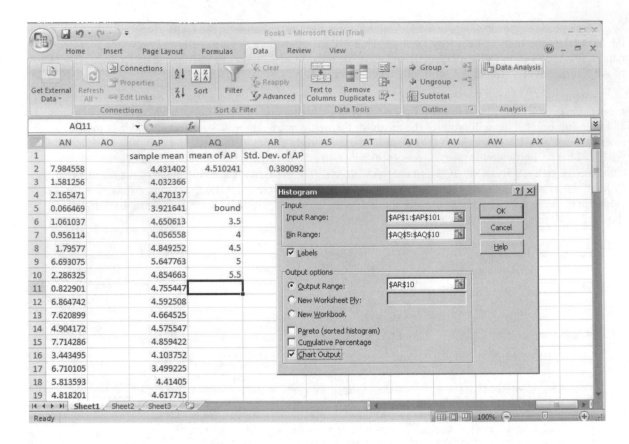

Click **OK.** Your results should be similar to those shown below.

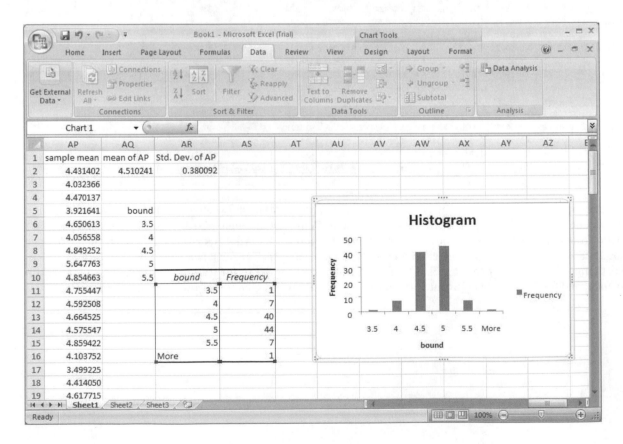

Copyright © Cengage Learning. All rights reserved.

CHAPTER 8: ESTIMATION

CONFIDENCE INTERVALS FOR THE MEAN – WHEN σ IS KNOWN (SECTION 8.1 OF *UNDERSTANDING BASIC STATISTICS*)

Excel's function for computing confidence intervals for a population mean assumes a normal distribution, regardless of sample size. Therefore, as stated in the text, if the population distribution is not a normal distribution, then a large sample should be used. The command syntax is

CONFIDENCE(alpha,standard_dev,size),

where alpha equals 1 minus the confidence level. In other words, an alpha of 0.05 indicates a 95% confidence interval. To generate a 99% confidence interval, use alpha = 0.01. The standard deviation is the population standard deviation σ.

Recall from the discussion in *Understanding Basic Statistics* that a confidence interval for the population mean has the form

$$\bar{x} - E \leq \mu \leq \bar{x} + E$$

Excel uses the formula

$$E = z_c \frac{\sigma}{\sqrt{n}}$$

where z_c is the critical value for the chosen confidence level. For $c = 95\%$, $z_c = 1.96$.

CONFIDENCE returns the value of E. To find the lower boundary of the confidence interval, you must subtract E from the sample mean $\bar{x}$ of your data; to find the upper boundary, you add E to the sample mean.

Example

Lucy decided to try to estimate the average number of miles she drives each day. For a three-month period, she selected a random sample of 35 days and kept a record of the distance driven on each of those sample days. The sample mean was 28.3 miles; assume that the standard deviation was $\sigma = 5$ miles. Find a 95% confidence interval for the population mean of miles Lucy drove per day in the three-month period.

For a 95% confidence level, alpha is 0.05. In Excel worksheet, after putting in some information and labels, we type =CONFIDENCE(0.05,5,35) in Cell B3 to represent E, and press Enter. In Cell D4, we use =28.3–B3 to compute the lower value of the confidence interval, and we compute the upper value of the confidence interval in Cell F4 using =28.3+B3.

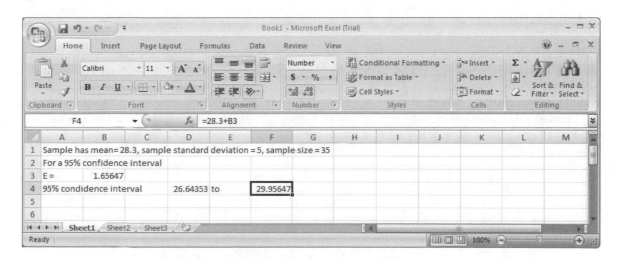

Copyright © Cengage Learning. All rights reserved.

CONFIDENCE INTERVALS FOR THE MEAN – WHEN σ IS UNKNOWN (SECTION 8.2 OF *UNDERSTANDING BASIC STATISTICS*)

When the population standard deviation σ is unknown, the sample standard deviation s is used to replace σ, and the confidence intervals are computed using Student's t distribution.

The confidence interval is computed as

$$\bar{x} - E \text{ to } \bar{x} + E$$

$$\text{where } E = t_c \frac{s}{\sqrt{n}}.$$

Excel has two commands for Student's t distribution under the statistical options listed in the Insert Function dialog box.

> **TDIST(x,deg_freedom,tails)**
>> returns the area in the tail of Student's t distribution beyond the specified value of x, for the specified number of degrees of freedom and number of tails (1 or 2).

> **TINV(probability,deg_freedom)**
>> returns the t_c value such that the area in the two tails beyond the t_c value equals the specified probability for the specified degrees of freedom.

We can use the TINV command to find the t_c value to use in the confidence interval. For instance, suppose we have a sample of size 10 with mean $\bar{x} = 6$ and $s = 1.2$. For a 95% confidence interval, 5% of the area lies beyond t_c in the two tails. The number of degrees of freedom is $10 - 1 = 9$. Therefore, from TINV we find that the t_c value for use in the computation of a 95% confidence interval is 2.262.

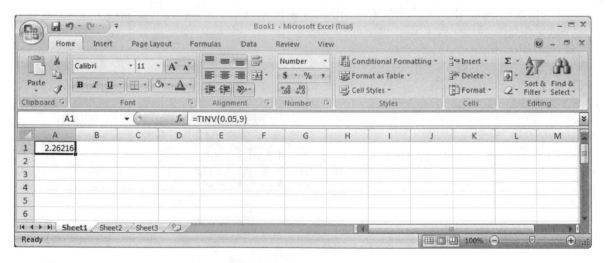

We can also find the E value in the Descriptive Statistics dialog box. Click on the **Data Analysis** button in the **Analysis** group under the **Data** tab to open the Data Analysis dialog box. In the **Analysis Tools** window, select **Descriptive Statistics,** and click **OK.** This opens the Descriptive Statistics dialog box. If you have data entered into a worksheet, you can use these menu selections to automatically compute the sample mean $\bar{x}$ and the sample standard deviation s for the data, as well as the value of E for the confidence interval on the basis of Student's t distribution, no matter what the sample size is.

Copyright © Cengage Learning. All rights reserved.

Example

The manager of First National Bank wishes to know the average waiting times for student loan application action. A random sample of 20 applications showed the waiting times from application submission (in days) to be

3	7	8	24	6	9	12	25	18	17
4	32	15	16	21	14	12	5	18	16

Find a 95% confidence interval for the population mean of waiting times.

Enter the data in Column A. Then widen cells in Columns C and D to accommodate the output. Follow the instructions on the previous page to open the Descriptive Statistics dialog box. Because the weights are in Column A, we select the cells containing the range and use these for the input range. We place the upper left corner of the output in Cell C1 and check that we want summary statistics and a 95% confidence interval.

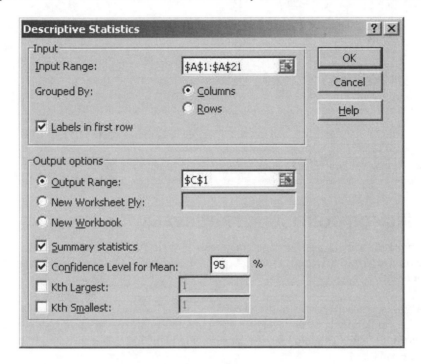

Click **OK**. Notice that the sample mean is in Cell D3 and that the value of E for a 95% confidence interval (again, based on Student's t distribution, not the standard normal distribution) is in Cell D16. We put the confidence interval in Cells F2 to H2 by entering =D3–D16 in Cell F2 and =D3+D16 in Cell H2. The display should be similar to the one shown at the top of the next page.

Copyright © Cengage Learning. All rights reserved.

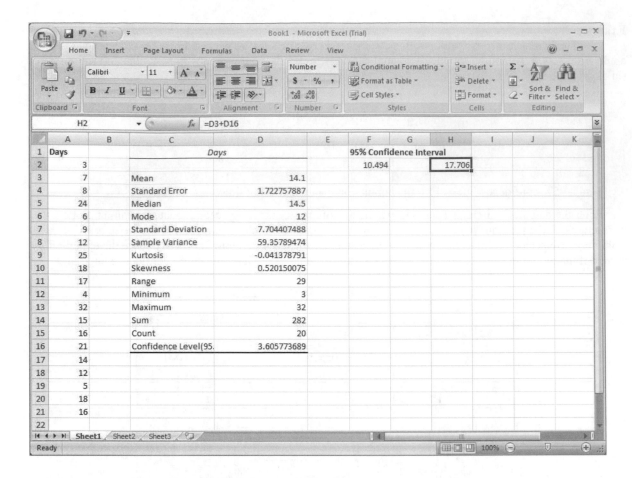

LAB ACTIVITIES FOR CONFIDENCE INTERVALS FOR THE MEAN

1. Retrieve the worksheet Sv03.xls. This contains the heights (in feet) of 65 randomly selected pro basketball players. Use the Descriptive Statistics dialog box to get summary statistics for the data and to create a 90% confidence interval.

2. Retrieve the worksheet Sv01.xls from the Excel data disk. This worksheet contains the number of shares of Disney Stock (in hundreds of shares) sold for a random sample of sixty trading days in 1993 and 1994. Use the Descriptive Statistics dialog box to get summary statistics for the data and to create the following confidence intervals.

   (a) Find a 99% confidence interval for the population mean volume.

   (b) Find a 95% confidence interval for the population mean volume.

   (c) Find a 90% confidence interval for the population mean volume.

   (d) Find an 85% confidence interval for the population mean volume.

   (e) What do you notice about the lengths of the intervals as the confidence level decreases?

3. There are many types of errors that will cause a computer program to terminate or give incorrect results. One type of error is punctuation. For instance, if a comma is inserted in the wrong place, the program might not run. A study of programs written by students in a beginning programming course showed that 75 out of 300 errors selected at random were punctuation errors. Find a 99% confidence interval for the proportion of errors made by beginning programming students that are punctuation errors. Next, find a 90% confidence interval. Use the CONFIDENCE command to find the interval. Is this second interval longer or shorter than the first?

Copyright © Cengage Learning. All rights reserved.

CHAPTER 9: HYPOTHESIS TESTING

TESTING A SINGLE POPULATION MEAN – WHEN σ IS KNOWN (SECTION 9.2 OF *UNDERSTANDING BASIC STATISTICS*)

Chapter 9 of *Understanding Basic Statistics* introduces tests of hypotheses. Testing involving a single mean is found in Section 9.2. Hypothesis tests in this section test the value of the population mean, μ, against some specified value, denoted by k.

When population standard deviation σ is known, one-sample z-tests are appropriate for testing the null hypothesis $H_0: \mu = k$ against one of the three alternative hypotheses $H_1: \mu > k$, $H_1: \mu < k$, or $H_1: \mu \neq k$ when (1) the data in the sample are known to be from a normal distribution (in which case any sample size will do) or when (2) the data distribution is unknown or the data are believed to be from a non-normal distribution, but the sample size, n, is large ($n \geq 30$).

In Excel, the ZTEST function finds the P-value for an upper- or right-tailed test, used to decide between the hypotheses $H_0: \mu = k$ and $H_1: \mu > k$. The null hypothesis says that the value of the population mean μ is k. A right-tailed test is used when the sample mean $\bar{x}$ is greater than k, suggesting that μ may in fact be greater than k, as the alternative hypothesis states.

The syntax for the ZTEST command is

ZTEST(array,x,sigma).

The array is the list of sample values; what Excel calls x is k, and sigma is the known value of σ, the population standard deviation. If the syntax used is **ZTEST(array,x),** i.e. if there is no sigma value given, then Excel calculates the sample standard deviation, s, from the sample data in the array and uses that in place of σ.

The P-value returned by ZTEST is the probability, given that the null hypothesis is true, of getting results at least as extreme as those observed in the sample. More precisely, ZTEST gives the probability of obtaining a sample mean greater than or equal to the observed sample mean, $\bar{x}$. When this probability is small, it means that the data in the observed sample would be surprising if H_0 were true. This is a reason to reject H_0.

ZTEST can also be used to apply a left-tailed test ($H_0: \mu = k$ versus $H_1: \mu < k$) or a two-tailed test ($H_0: \mu = k$ versus $H_1: \mu \neq k$). To apply a left-tailed test, for the case where the sample mean $\bar{x}$ is less than k, simply apply a right-tailed test and then subtract the result from 1. (When $\bar{x} < k$, the area found by ZTEST in the upper "tail" will be greater than 0.5, and 1 minus that area will be the area in the lower tail.) To apply a two-tailed test, either double the P-value from a right-tailed test (when $\bar{x} > k$) or double the P-value from a left-tailed test (when $\bar{x} < k$).

To call up the ZTEST dialog box, click the **Insert Function** button on the left side of the Formula bar, select Statistical in the **Or select a category** window, scroll down in the **Select a function** window until you reach ZTEST, and click **OK.** The dialog box should be similar to the one shown on the top of the next page.

Copyright © Cengage Learning. All rights reserved.

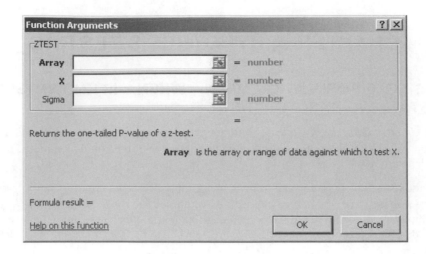

Enter the cell range containing the sample values in the **Array** window, and in the window for **X** enter the mean given by the null hypothesis. In the **Sigma** window, enter the value of the population standard deviation σ if it is known. Again, sigma is optional; if this box is left blank, Excel will compute the sample standard deviation for the data in the specified array and use s instead of σ in the computation for z. Recall that if we are dealing with large samples, s and σ are fairly close, so this approximation produces reliable results. Finally, when you click on **OK,** the P-value of the right-tailed test for $\bar{x}$ is computed.

You can also type the command =ZTEST(array,x,sigma) directly into the Formula bar and press Enter.

Once the P-value has been computed, we can then compare it with α, the level of significance of the test. If P-value $\le \alpha$, we reject the null hypothesis. If P-value $> \alpha$, we do not reject the null hypothesis.

Example

ZTEST requires the use of large samples (size 30 or greater) when the population distribution is unknown. Let us consider the following data, which contains heights, in feet, of 32 randomly selected professional basketball players.

6.5	6.25	6.33	6.55	6.4	6.37	6.8	6.23
7.1	5.9	6.45	6.7	6.55	6.4	6.38	6.2
6.72	6.9	6.5	6.44	6.8	5.9	6.0	6.9
6.3	7.2	6.3	6.82	6.32	6.44	6.56	6.71

Assume that twenty years ago, the average height of professional basketball players was 6.3 feet (that translates to 6 feet, 3.6 inches). Let's use the above data to consider whether the current population mean height of professional basketball players is greater than it was twenty years ago. The null hypothesis will be that their average height is the same. Given our alternative hypothesis ("greater than"), we will apply a right-tailed test.

Enter the data in Column A. Cell A1 contains the label Heights. Enter the data in Cells A2:A33.

After typing in some labeling information, we want to display the P-value provided by ZTEST in Cell F3. Activate Cell F3 and, in the Insert Function dialog box, select Statistical and ZTEST. The Function Arguments dialog box should be similar to the one shown at the top of the next page.

We will use the cell range A2:A33 for **Array** and 6.3 as the value for **X.** The population standard deviation is not given in this case. To demonstrate the use of ZTEST, let's use the value of s, the sample standard deviation, for the value of σ, and let Excel compute s from the sample data. Notice that with the space for **Sigma** left blank, the dialog box tells us that the P-value is about 0.00016. We interpret this as the probability that 32 data values could come out with a mean greater than or equal to that of the sample, given that they were taken from a normal distribution with a mean of 6.3.

Copyright © Cengage Learning. All rights reserved.

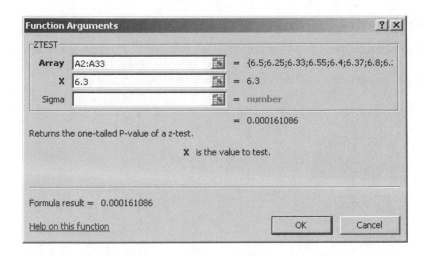

Click **OK.** Because the *P*-value is less than even the very restrictive $\alpha = 0.01$, we reject the null hypothesis and conclude that the population mean height of professional basketball players now is greater than it was twenty years ago.

For completeness, we opened the Descriptive Statistics dialog box and generated descriptive statistics for our data. First we widened Column C to fit the display, clicked the **Data Analysis** button in the **Analysis** group under the **Data** tab, and selected **Descriptive Statistics** in the Data Analysis dialog box. We then typed $A\$1:\$A\$33 for the Input Range (grouped by columns with labels in the first row) and $C\$5 for the Output Range, checked Summary statistics, typed 95% for the Confidence Level for Mean, and clicked OK.

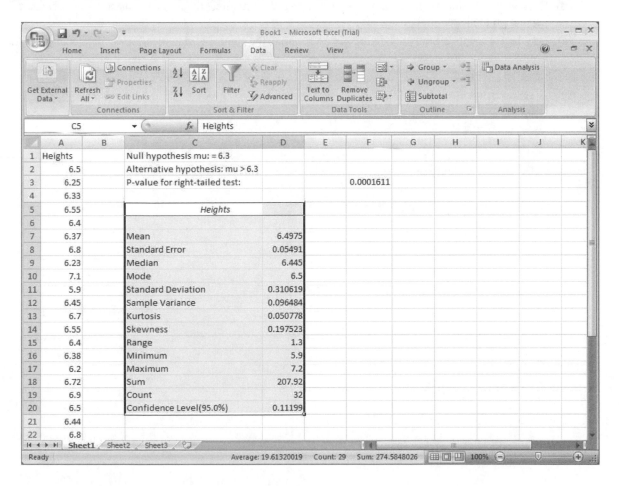

Copyright © Cengage Learning. All rights reserved.

LAB ACTIVITIES FOR TESTING A SINGLE POPULATION MEAN

1. Open or retrieve the worksheet Sv04.xls from the student website. The data in Column A of this worksheet represent the miles-per-gallon gasoline consumption (highway) for a random sample of 55 makes and models of passenger cars (source: Environmental Protection Agency).

30	27	22	25	24	25	24	15
35	35	33	52	49	10	27	18
20	23	24	25	30	24	24	24
18	20	25	27	24	32	29	27
24	27	26	25	24	28	33	30
13	13	21	28	37	35	32	33
29	31	28	28	25	29	31	

Test the hypothesis that the population mean mile-per-gallon gasoline consumption for such cars is greater than 25 mpg.

(a) Do we know σ for the mpg consumption? If not, use the value of s for the value of σ (sometimes this is done in practice when sample size is large.) Can we use the normal distribution for the hypothesis test?

(b) State the null and alternate hypothesis and type them on your worksheet.

(c) Use ZTEST with Sigma omitted.

(d) Look at the P-value in the output. Compare it to α. Do we reject the null hypothesis or not? Does it depend on the level of significance?

(e) Use the Descriptive Statistics dialog box to generate the summary statistics for the data, and place the results on the worksheet.

Copyright © Cengage Learning. All rights reserved.

CHAPTER 10: INFERENCES ABOUT DIFFERENCES

TESTS INVOLVING PAIRED DIFFERENCES – DEPENDENT SAMPLES (SECTION 10.1 OF *UNDERSTANDING BASIC STATISTICS*)

The test for difference of means of dependent samples is presented in Section 10.1 of *Understanding Basic Statistics*. Dependent samples arise from before-and-after studies, some studies of data taken from the same subjects, and some studies on identical twins.

In Excel there are two functions that produce the *P*-value for a one- or two-tailed test of paired differences. The first command is TTEST, found by opening the Insert Function dialog box and selecting Statistical and TTEST. You can also activate a cell and type the command with correct arguments in the Formula bar. This command returns only the *P*-value for the test. The syntax is

TTEST(array1,array2,tails,type)

If tails = 1, then TTEST returns the *P*-value for a one-tailed test, and if tails = 2, then TTEST returns the two-tailed value. For the parameter called type, there are three choices:

type	Test performed using Student's *t* distribution
1	Paired difference test
2	Difference of means test for two samples with equal variances
3	Difference of means test for two samples with unequal variances

The other Excel function can be found in the **Analysis** group under the **Data** tab. In the Data Analysis dialog box, the **t-Test: Paired Two Sample for Means** option provides much more information than the TTEST function. We will use this alternative approach in the next example.

Example

Promoters of a state lottery decided to advertise the lottery heavily on television for one week during the middle of one of the lottery games. To see if the advertising improved ticket sales, they surveyed a random sample of 8 ticket outlets and recorded weekly sales for one week before the television campaign and for one week after the campaign. The results (in ticket sales) follow, where Row A gives sales prior to the campaign and Row B gives sales afterward.

A	3201	4529	1425	1272	1784	1733	2563	3129
B	3762	4851	1202	1131	2172	1802	2492	3151

Test the claim that the television campaign increased lottery ticket sales at the 0.05 level of significance.

Enter the data in Columns A and B, with appropriate headers, widen the output column (Column D) to fit the display. Next, open the dialog box as shown on the next page by clicking the **Data Analysis** button in the **Analysis** group, highlighting **t-Test: Paired Two Sample for Means** in the **Analysis Tools** window, and clicking **OK.**

Copyright © Cengage Learning. All rights reserved.

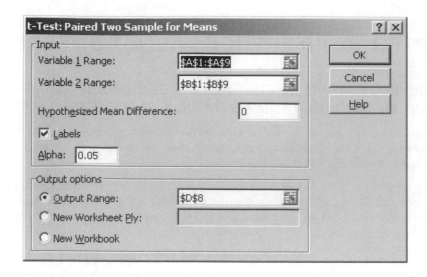

Notice that we use Column A cells for **Variable 1 Range,** Column B cells for **Variable 2 Range,** and we check the **Labels** box. The null hypothesis is H_o: $\mu = 0$, so we enter 0 as the value for the **Hypothesized Mean Difference.** We type 0.05 for **Alpha,** select Cell D8 as the upper left cell for the **Output Range,** and click **OK.**

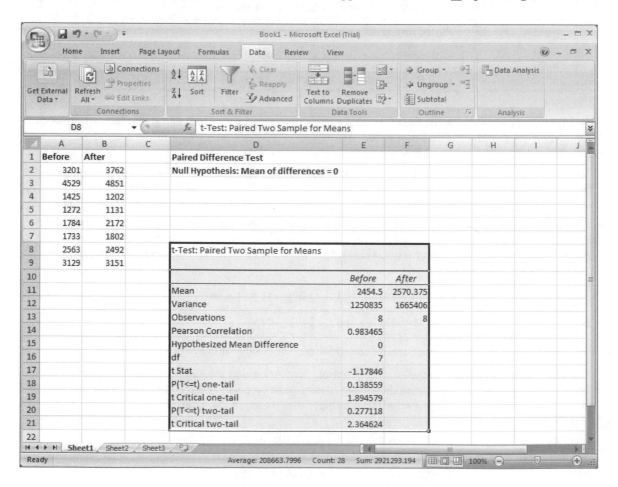

Copyright © Cengage Learning. All rights reserved.

Notice that we get $P = 0.1386$ for a one-tailed test. Because this value is larger than the level of significance, we do not reject the null hypothesis. The same output gives the P-value for a two-tailed test as well. In addition, we see the sample t value of -1.17846, together with the critical values for a one- or two-tailed test using $\alpha = 0.05$.

LAB ACTIVITIES FOR TESTS INVOLVING PAIRED DIFFERENCES

1. Open or retrieve the worksheet Tvds01.xls from the student website. The data are pairs of values where the entries in Column A represents average salary ($1000/yr) for male faculty members at an institution and those in Column B represent the average salary for female faculty members ($1000/yr) at the same institution. A random sample of 22 U.S. colleges and universities was used (source: *Academe*, Bulletin of the American Association of University Professors).

(34.5, 33.9)	(30.5, 31.2)	(35.1, 35.0)	(35.7, 34.2)	(31.5, 32.4)
(34.4, 34.1)	(32.1, 32.7)	(30.7, 29.9)	(33.7, 31.2)	(35.3, 35.5)
(30.7, 30.2)	(34.2, 34.8)	(39.6, 38.7)	(30.5, 30.0)	(33.8, 33.8)
(31.7, 32.4)	(32.8, 31.7)	(38.5, 38.9)	(40.5, 41.2)	(25.3, 25.5)
(28.6, 28.0)	(35.8, 35.1)			

   (a) Enter the data in Columns A and B of the worksheet.

   (b) Select **t-Test: Paired Two Sample for Means** in the Data Analysis dialog box to test the hypothesis that there is a difference in salary. What is the P-value of the sample test statistic? Do we reject or fail to reject the null hypothesis at the 5% level of significance? What about at the 1% level of significance?

   (c) Select **t-Test: Paired Two Sample for Means** in the Data Analysis dialog box to test the hypothesis that female faculty members have a lower average salary than male faculty members. What is the test conclusion at the 5% level of significance? At the 1% level of significance?

2. An audiologist is conducting a study on noise and stress. Twelve subjects selected at random were given a stress test in a room that was quiet. Then the same subjects were given another stress test, this time in a room with high-pitched background noise. The results of the stress tests were scores 1 through 20 with 20 indicating the greatest stress. The results follow, where A represents the scores of the test administered in the quiet room and B represents the scores of the test administered in the room with the high-pitched background noise.

Subject	1	2	4	5	6	7	8	9	10	11	12
A	13	12	16	19	7	13	9	15	17	6	14
B	18	15	14	18	10	12	11	14	17	8	16

   Test the hypothesis that the stress level was greater during exposure to high-pitched background noise. Look at the P-value. Should you reject the null hypothesis at the 1% level of significance? At the 5% level?

Copyright © Cengage Learning. All rights reserved.

TESTING OF DIFFERENCES OF MEANS (SECTION 10.2 OF *UNDERSTANDING BASIC STATISTICS*)

Tests of difference of means for independent samples are presented in Section 10.2 of *Understanding Basic Statistics*. We consider the $\bar{x}_1 - \bar{x}_2$ distribution. The null hypothesis is that there is no difference between means, so $H_0: \mu_1 = \mu_2$ or $H_0: \mu_1 - \mu_2 = 0$.

When σ_1 and σ_2 are known

When testing the difference of means with known σ_1 and σ_2, the z-test based on a normal distribution is used if either the population has a normal distribution or the sample size is large. The values of population variance, which is the square of population standard deviation, are actually needed in the Excel testing procedure. Some examples used in this section do not provide the values of population standard deviations.

To demonstrate the testing procedure, we use the sample variance values for the population variance values. In Excel, we select VAR from the Insert Function dialog box and enter the data range. Then we use **z-Test: Two Sample for Means** in the Data Analysis dialog box to obtain the sample z statistic, *P*-values for a one- or two-tailed test, and critical z values for one- or two-tailed tests at the specified level of significance.

Example

The following data contain heights, in feet, of 36 randomly selected non-professional basketball players. Let us compare these data with the heights of 32 professional basketball players from the previous example.

6.1	6.2	6.21	6.5	6.1	5.7	5.8	6.3	5.7
6.1	5.7	6.5	5.7	6.3	6.4	6.3	6.0	5.9
5.72	5.9	6.1	5.44	5.8	5.9	6.0	5.9	5.5
6.1	6.2	6.3	5.8	6.2	6.4	5.8	5.71	6.0

Re-enter the data from the previous example in Column A under the heading "Pro." Enter the above data in Column B under the heading "Non-Pro," and widen the Column D to fit the output display. Calculate the sample variances using the VAR command for the data in Columns A and B, respectively, and place the results in Cells E1 and E2. Then open the Data Analysis dialog box, select **z-Test: Two Sample for Means** in the **Analysis Tools** window, and click **OK.** Enter the values in the resulting dialog box as shown below.

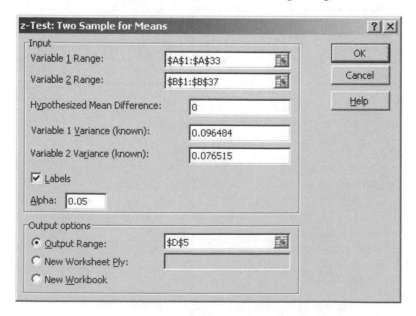

Copyright © Cengage Learning. All rights reserved.

Click **OK.** The result follows.

When σ_1 and σ_2 are unknown

To do a test of difference of sample means when σ_1 and σ_2 are unknown, *t*-tests are used if either the population has approximately a normal distribution or the sample size is large. If we assume that the samples come from populations with the same standard deviation, we then use **t-Test: Two-Sample Assuming Equal Variances** analysis tool in the Data Analysis dialog box. If we believe that population standard deviations are not equal, we use the **t-Test: Two-Sample Assuming Unequal Variances** analysis tool.

Example

Sellers of microwave French fry cookers claim that their process saves cooking time. The McDougal Fast Food Chain is considering the purchase of these new cookers, but wants to test the claim.

Six batches of French fries were cooked in the traditional way. The cooking times (in minutes) were

> 15   17   14   15   16   13

Six batches of French fries of the same weight were cooked using the new microwave cooker. These cooking times (in minutes) were

> 11   14   12   10   11   15

Let us assume that both populations are approximately normal with equal standard deviations. Test the claim that the microwave process takes less time. Use $\alpha = 0.05$.

Copyright © Cengage Learning. All rights reserved.

Enter the traditional data in Column A, the new data in Column B, and widen Column D to fit the output display. Then open the Data Analysis dialog box in the **Analysis** group, select **t-Test: Two-Sample Assuming Equal Variances** in the **Analysis Tools** window, and click **OK.** Enter the values in the resulting dialog box as shown below. Note that the hypothesized mean difference is zero.

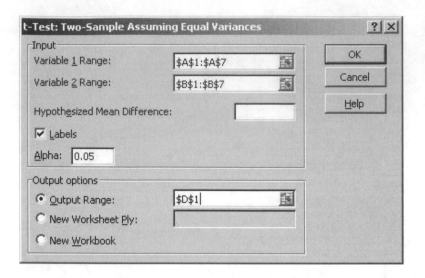

Click **OK.** The result follows.

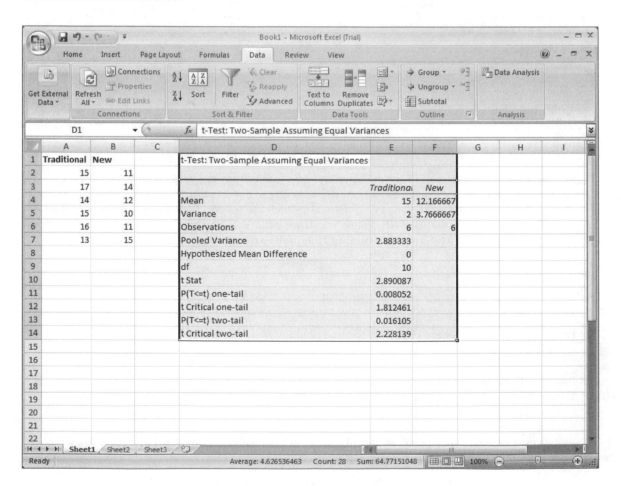

Copyright © Cengage Learning. All rights reserved.

We see that the *P*-value for a one-tail test is 0.00805. Because this value is less than 0.05, we reject the null hypothesis and conclude that the new method takes less time on average.

LAB ACTIVITIES FOR TESTING DIFFERENCES OF MEANS

1. Calm Cough Medicine is testing a new ingredient to see if its addition will lengthen the effective cough relief time of a single dose. A random sample of fifteen doses of the standard medicine was tested, and the effective relief times (in minutes) were

   42   35   40   32   30   26   51   39   33   28

   37   22   36   33   41

   Then a random sample of twenty doses with the new ingredient was tested. The effective relief times (in minutes) were

   43   51   35   49   32   29   42   38   45   74

   31   31   46   36   33   45   30   32   41   25

   Assume that the standard deviations of the relief times are equal for the two populations. Also assume that both populations are approximately normal. Test the claim that the effective relief time is longer when the new ingredient is added. Use $\alpha = 0.01$.

2. Retrieve the worksheet Tvis06.xls from the student website. The data represent numbers of cases of red fox rabies for a random sample of sixteen areas in each of two different regions of southern Germany.

   Number of Cases in Region 1

   10   2   2   5   3   4   3   3   4   0   2   6   4   8   7   4

   Number of Cases in Region 2

   1   1   2   1   3   9   2   2   4   5   4   2   2   0   0   2

   Test the hypothesis that the average number of cases in Region 1 is greater than the average number of cases in Region 2. Use a 1% level of significance. Assume that both populations are approximately normal and have equal standard deviations.

3. Retrieve the Excel worksheet Tvis02.xls from the student website. The data represent the petal length (in centimeters) for a random sample of 35 *Iris Virginica* plants and for a random sample of 38 *Iris Setosa* plants (source: Anderson, E., Bulletin of American Iris Society).

   Petal Length of *Iris Virginica*

   5.1   5.8   6.3   6.1   5.1   5.5   5.3   5.5   6.9   5.0   4.9   6.0   4.8   6.1   5.6   5.1

   5.6   4.8   5.4   5.1   5.1   5.9   5.2   5.7   5.4   4.5   6.1   5.3   5.5   6.7   5.7   4.9

   4.8   5.8   5.1

   Petal Length of *Iris Setosa*

   1.5   1.7   1.4   1.5   1.5   1.6   1.4   1.1   1.2   1.4   1.7   1.0   1.7   1.9   1.6   1.4

   1.5   1.4   1.2   1.3   1.5   1.3   1.6   1.9   1.4   1.6   1.5   1.4   1.6   1.2   1.9   1.5

   1.6   1.4   1.3   1.7   1.5   1.7

   Test the hypothesis that the average petal length for *Iris Setosa* is shorter than the average petal length for *Iris Virginica*. Assume that population standard deviations are unequal.

Copyright © Cengage Learning. All rights reserved.

CHAPTER 11: ADDITIONAL TOPICS USING INFERENCE

CHI-SQUARE TEST OF INDEPENDENCE (SECTION 11.1 OF *UNDERSTANDING BASIC STATISTICS*)

Use of the chi-square distribution to test independence is discussed in Section 11.1 of *Understanding Basic Statistics*. In such tests we use hypotheses

H_0: The variables are independent

H_1: The variables are not independent

In Excel, the applicable command syntax is

CHITEST(actual_range,expected_range),

which returns the *P*-value of the sample χ^2 value, where the sample χ^2 value is computed as

$$\chi^2 = \sum \frac{(O-E)^2}{E}$$

Here, *E* stands for the expected count in a cell, and *O* stands for the observed count in that same cell. The sum is taken over all cells.

In our Excel worksheet, we first enter the contingency table of observed values. If the table does not contain column sums or row sums, use the Sum button on the tool bar to generate the sums. We need to create the table of expected values, where the expected value *E* for a cell is

$$E = (\text{column total})\left(\frac{\text{row total}}{\text{grand total}}\right)$$

By careful use of absolute and relative cell references, we can type the formula once and then copy it to different positions in the contingency table of expected values. Recall that an absolute cell reference has \$ symbols preceding the column and row designators.

Example

A computer programming aptitude test has been developed for high school seniors. The test designers claim that scores on the test are independent of the type of school the student attends: rural, suburban, urban. A study involving a random sample of students from these types of institutions yields the following contingency table. Use the CHITEST command to compute the *P*-value of the sample chi-square value. Then determine if the type of school and test score are independent at the $\alpha = 0.05$ level of significance.

	School Type		
Score	Rural	Suburban	Urban
200-299	33	65	83
300-399	45	79	95
400-500	21	47	63

First we enter the table into a worksheet and use the **Σ ▾** button (**Sum**) in the **Editing** group under the **Home** tab to generate the column, row, and grand total sums.

Copyright © Cengage Learning. All rights reserved.

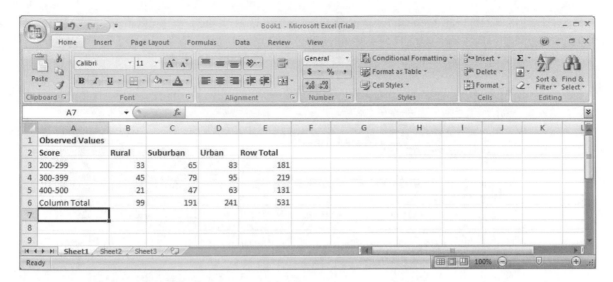

Next we create the contingency table of expected values, where the expected value for Cell B3 will go in Cell H3. Notice that in the formula

$$E = (\text{row total})\left(\frac{\text{row total}}{\text{grand total}}\right)$$

the grand total stays the same for each expected value. The grand total is in Cell E6. Because we want this to be an absolute address used in each computation, we use the cell label $E$6. (Alternatively, we could just type in the grand total of 531 in the Formula bar.) The column totals are all in Row 6, so when we refer to a column total, we will fix the row address by using $6 and let the column names vary. The row totals are all in Column E, so we will fix the column address as $E and let the row address vary when we use row totals. So the formula as entered in Cell H3 should be

$$=B\$6*(\$E3/\$E\$6)$$

Now move the cursor to the lower right corner of Cell H3. When the small + appears, press down on the left mouse button and drag the + to the lower right corner of Cell J3. Release the mouse button. Then move the cursor to the lower right corner of the selected Cells H3 through J3, and when the small + appears, press the mouse button and drag the + to the lower right corner of Cell J5. The calculations for all the cells will automatically be done. Adjust the values for five decimal places.

We now have both the observed values and the expected values.

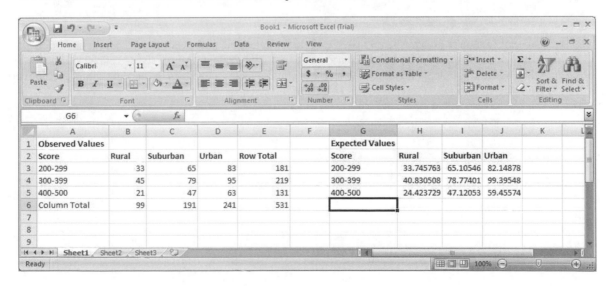

Copyright © Cengage Learning. All rights reserved.

Now use the CHITEST command. Open the Insert Function dialog box, select Statistical in the **Or select a category** window, scroll down in the **Select a function** window until you reach CHITEST, and click **OK.** The following Function Arguments dialog box will appear. Enter the ranges of values as shown below.

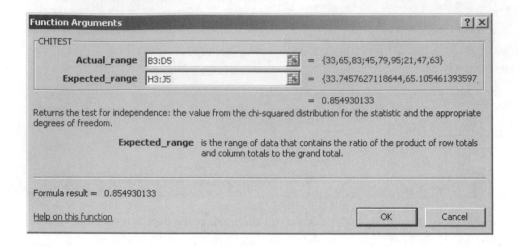

Click **OK.** The resulting *P*-value is 0.8549. Place appropriate labels on the worksheet.

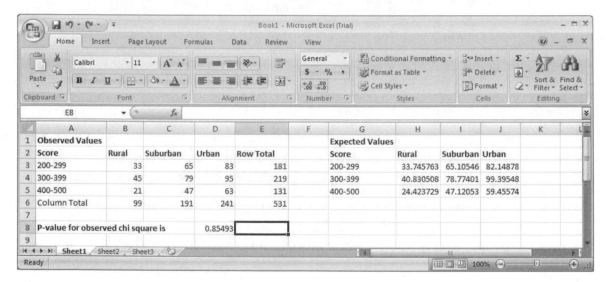

Because the *P*-value is greater than $\alpha = 0.05$, we do not reject the null hypothesis. There is insufficient evidence to conclude that school type and test scores are not independent. Intuitively, we can confirm this by observing that, assuming independence, the values we would expect (the values in the right table) are not very different from the values observed (the values in the left table).

Copyright © Cengage Learning. All rights reserved.

LAB ACTIVITIES FOR CHI-SQUARE TEST OF INDEPENDENCE

In each activity, enter the contingency tables into a worksheet, then use the Sum button to generate the required row and column sums. Create a table of expected values. Finally, use CHITEST to find the P-value of the sample statistic and draw the appropriate conclusion.

1. We Care Auto Insurance had its staff of actuaries conduct a study to see if vehicle type and loss claim are independent. A random sample of auto claims over the first six months gives the information in this contingency table.

Type of Vehicle	Total Loss Claims per Year per Vehicle			
	$0–999	$1000–2999	$3000–5999	$6000+
Sports Car	20	10	16	8
Truck	16	25	33	9
Family Sedan	40	68	17	7
Compact	52	73	48	12

Test the claim that car type and loss claim are independent. Use $\alpha = 0.05$.

2. An educational specialist is interested in comparing three methods of instruction:

SL – standard lecture with discussion

TV – video-taped lectures with no discussion

IM – individualized method with reading assignments and tutoring, but no lectures

The specialist conducted a study of these three methods to see if they were independent. A course was taught using each of the three methods, and a standard final exam given at the end. Students were put into the different-method sections at random. The course type and test results are shown in the contingency table below.

Course Type	Final Exam Score				
	<60	60–69	70–79	80–89	90–100
SL	10	4	70	31	25
TV	8	3	62	27	23
IM	7	2	58	25	22

Test the claim that the instruction method and final exam test scores are independent, using $\alpha = 0.01$.

Note: If you have raw data entered in columns, you can click on one of the cells, then on the **PivotTable** button in the **Tables** group under the **Insert** tab to construct a contingency table. See the Excel help menu or other manuals for details.

Copyright © Cengage Learning. All rights reserved.

TABLE OF EXCEL FUNCTIONS

DESCRIPTIVE STATISTICS

These functions can be entered by using the **Insert Function** button to open the Insert Function dialog box, selecting Statistical in the **Or select a <u>c</u>ategory** window, and then highlighting the function name in the **Select a functio<u>n</u>** window. The function arguments are then entered in the dialog boxes that open subsequently. Alternatively, you can type =, followed by the command name and arguments directly in the Formula bar.

=AVERAGE(number1,number2,...)
returns the average (arithmetic mean) of its arguments, which can be numbers or names, arrays, or references that contain numbers.

=COUNT(value1,value2,...)
counts the number of cells in a range that contain numbers.

=COUNTIF(range,criteria)
counts the number of cells within a range that meet the given condition.

=MAX(number1,number2,...)
returns the largest value in a set of values. Ignores logical values and text.

=MEDIAN(number1,number2,...)
returns the median, or the number in the middle of the set of given numbers.

=MIN(number1,number2,...)
returns the smallest number in a set of values. Ignores logical values and text.

=MODE(number1,number2,...)
returns the most frequently occurring, or repetitive, value in an array or range of data.

=QUARTILE(array,quart)
returns the quartile of a data set.

=STDEV(number1,number2,...)
estimates standard deviation based on a sample (ignores logical values and text in the sample).

=STDEVP(number1,number2,...)
calculates standard deviation based on the entire population given as arguments (ignores logical values and text).

=TRIMMEAN(array,percent)
returns the mean of the interior portion of a set of data values.

=VAR(number1,number2,...)
estimates variance based on a sample (ignores logical values and text in the sample).

=VARP(number1,number2,...)
calculates variance based on the entire population, including logical values and text. Text and the logical value FALSE have the value 0; the logical value TRUE has the value 1.

Clicking the **Data Analysis** button in the **Analysis** group under the **Data** tab opens the Data Analysis dialog box. Selecting **Descriptive Statistics** in the **<u>A</u>nalysis Tools** window and clicking **OK** opens the Descriptive Statistics dialog box. For a given range of values entered as the **<u>I</u>nput Range,** checking **<u>S</u>ummary statistics** and clicking **OK** returns values for the mean, standard error of the mean, median, mode, standard deviation, variance, kurtosis, skewness, range, minimum, maximum, sum, and count of the data. Checking and entering a percentage value for **Co<u>n</u>fidence Level for Mean** and then clicking **OK** returns the error E for a confidence interval $\bar{x} - E$ to $\bar{x} + E$ at the designated percentage confidence level.

          Copyright © Cengage Learning. All rights reserved.

GRAPHICS

To create a histogram in Excel, click on the **Data Analysis** button in the **Analysis** group under the **Data** tab. This opens the Data Analysis dialog box. In the **Analysis Tools** window, select **Histogram** and click **OK.** This will open the Histogram dialog box, in which the desired input and output information for the histogram can be specified.

To create graphs for column charts (bar graphs), pie charts (circle graphs), line graphs (for time series), or scatter diagrams, select the desired graph type in the **Charts** group under the **Insert** tab. To obtain a regression line and formula for data in a scatter diagram, right-click on one of the data points and. select the **Add Trendline...** option in the drop-down menu. Then in the Format Trendline dialog box, select **Linear** as the **Trend/RegressionType,** check both **Display Equation on chart** and **Display R-squared value on chart** at the bottom of the dialog box, and click **Close.**

LINEAR REGRESSION

These functions can be entered by using the **Insert Function** button to open the Insert Function dialog box, selecting Statistical in the **Or select a category** window, and then highlighting the function name in the **Select a function** window. The function arguments are then entered in the dialog boxes that open subsequently. Alternatively, you can type =, followed by the command name and arguments directly in the Formula bar.

= CORREL(array1,array2)
returns the correlation coefficient, r, between two data sets.

=FORECAST(x,known_y's,known_x's)
returns the predicted y value for the specified x value, using extrapolation from the given pairs of x and y values. Note that you need to use a new FORECAST command for each different x value.

=INTERCEPT(known_y's,known_x's)
returns the intercept, a, of the least-squares line.

=LINEST(known_y's,known_x's,const,stats)
returns the slope b and y-intercept a of the least-squares line. (NOTE: this command can be found in the Insert Function dialog box, but it is better to type it in because it is an array formula. To use LINEST,

1. Activate two cells, the first to hold the slope b, the second for the intercept a.
2. Type =LINEST(known_y'x,known_x's) in the Formula bar, with the appropriate cell ranges in place of the known x and known y values.
3. Press **Ctrl+Shift+Enter**. This key combination activates the array formula features so that you get the outputs for both b and a. Otherwise, you will get only the slope b of the least-squares line.

=PEARSON(array1,array2)
returns the Pearson product moment correlation, r, for the two data sets.

=RSQ(known_y's,known_x's)
returns the square of the Pearson product moment correlation coefficient, r^2, through the given data points.

=SLOPE(known_y's,known_x's)
returns the slope, b, of the regression line through the given data points.

=STEYX(known_y's,known_x's)
returns the standard error of estimate, S_e, between the y values of the data and the predicted y values of the regression line.

Clicking the **Data Analysis** button in the **Analysis** group under the **Data** tab opens the Data Analysis dialog box. Selecting the **Regression** tool in the **Analysis Tools** window and clicking **OK** opens the Regression dialog box. The regression tool returns the coefficients for a simple regression model and provides for a table of residuals, a residual plot, and a normal probability plot.

Copyright © Cengage Learning. All rights reserved.

PROBABILITY

These functions can be entered by using the **Insert Function** button to open the Insert Function dialog box, selecting Statistical in the **Or select a category** window, and then highlighting the function name in the **Select a function** window. The function arguments are then entered in the dialog boxes that open subsequently. Alternatively, you can type =, followed by the command name and arguments directly in the Formula bar.

=BINOMDIST(number_s,trials,probability_s,cumulative)
returns the individual term binomial distribution probability. Using TRUE for cumulative returns the cumulative probability of obtaining no more than r successes in n trials. Using FALSE returns the probability of obtaining exactly r successes in n trials.

= CHIDIST(x,deg_freedom)
returns the one-tailed probability of the chi-square distribution with the designated degrees of freedom.

= CHIINV(probability,deg_freedom)
returns the inverse (x) of the one-tailed probability of the chi-square distribution.

=NORMDIST(x,mean,standard_dev,cumulative)
returns the normal cumulative distribution probability for the specified mean and standard deviation. Using TRUE for cumulative returns the cumulative probability that a random value $x \leq x_1$. Using FALSE returns the height of the normal probability density function evaluated at x_1.

=NORMINV(probability,mean,standard_dev)
returns the inverse (x) of the normal cumulative distribution for the specified mean and standard deviation. The area to the left of x is equal to the designated probability.

=NORMSDIST(z)
returns the standard normal cumulative distribution probability (i.e., a mean of 0 and a standard deviation of 1) for a randomly selected z score. This command is equivalent to NORMDIST(x,0,1,TRUE), which returns the probability that $z \leq z_1$.

=NORMSINV(probability)
returns the inverse (z) of the standard normal cumulative distribution (i.e., a mean of 0 and a standard deviation of 1). The area to the left of z is equal to the designated probability. This command is equivalent to NORMINV(probability,0,1).

=STANDARDIZE(x,mean,standard_dev)
returns the z score for the given x value from a distribution with the specified mean and standard deviation.

=TDIST(x,deg_freedom,tails)
returns the area in the tail of Student's t distribution beyond the specified value of x, for the specified number of degrees of freedom and number of tails (1 or 2). When tails is 1, the returned area is in the tail to the right of x. When tails is 2, the total area in both tails (to the right of x and to the left of $-x$) is returned.

=TINV(probability,deg_freedom)
returns the inverse (t_c) value of Student's t distribution, such that the area in the two tails beyond the t_c value equals the specified probability for the specified degrees of freedom.

Copyright © Cengage Learning. All rights reserved.

RANDOM SAMPLES

These functions can be entered by using the **Insert Function** button to open the Insert Function dialog box, selecting Math & Trig in the **Or select a category** window, and then highlighting the function name in the **Select a function** window. The function arguments are then entered in the dialog boxes that open subsequently. Alternatively, you can type =, followed by the command name and arguments directly in the Formula bar.

=RAND()
returns a random number greater than or equal to 0 and less than 1, evenly distributed (changes on recalculation).

=RANDBETWEEN(bottom,top)
returns a random number between the numbers you specify.

Clicking the **Data Analysis** button in the **Analysis** group under the **Data** tab opens the Data Analysis dialog box. Selecting **Random Number Generation** in the **Analysis Tools** window and clicking **OK** opens the Random Number Generation dialog box. Entering a given number of variables and random numbers, then selecting a distribution (uniform, normal, binomial, or discrete) and entering the appropriate parameters, and clicking **OK** returns a random sample.

Alternatively, by selecting **Sampling** in the **Analysis Tools** window and clicking **OK** opens the Sampling dialog box. Specifying the **Input Range,** checking **Random,** entering the number of random values desired in the **Number of Samples** window, and clicking **OK** returns a random sample from the designated cell range.

CONFIDENCE INTERVAL

Using the **Insert Function** button to open the Insert Function dialog box, select Statistical in the **Or select a category** window, highlight the function in the **Select a function** window. The function arguments are then entered in the dialog box that opens subsequently.

=CONFIDENCE(alpha,standard_dev,size)
returns the value of E in the confidence interval $\bar{x} - E$ to $\bar{x} + E$ for a population mean μ. The confidence level equals $100(1-\alpha)\%$.

Alternatively, you can type =, followed by the command name and arguments directly in the Formula bar.

Confidence levels for some of the other parameters are included in some of the outputs for certain dialog boxes that can be opened using the Data Analysis dialog box. One example of this has been mentioned in the previous Descriptive Statistics section.

Copyright © Cengage Learning. All rights reserved.

HYPOTHESIS TESTING

These functions can be entered by using the **Insert Function** button to open the Insert Function dialog box, selecting Statistical in the **Or select a category** window, and then highlighting the function name in the **Select a function** window. The function arguments are then entered in the dialog boxes that open subsequently. Alternatively, you can type =, followed by the command name and arguments directly in the Formula bar.

=CHITEST(actual_range,expected_range)
returns the P-value of the sample chi-square statistics for a test of independence.

=TTEST(array1,array2,tails,type)
returns the P-value of a Students t distribution for a one-tailed test if the value of tails is 1, and for a two-tailed test if the value of tails is 2. The type of test performed is as follows:

type = 1: a paired difference test is performed.
type = 2: a difference of means test for two samples with equal variance is performed.
type = 2: a difference of means test for two samples with unequal variance is performed.

=ZTEST(array,x,sigma)
returns the P-value for a right-tailed test, where H_0: $\mu = x$ is the null hypothesis and H_1: $\mu > x$ is the alternative hypothesis. To apply a left-tailed test (H_0: $\mu = x$ versus H_1: $\mu < x$), apply ZTEST and subtract the result from 1. To apply a two-tailed test, either double the P-value from a right-tailed test or double the P-value from a left-tailed test. For sigma, use the population standard deviation, if it is known. Otherwise leave the value for sigma out, and Excel will use the sample standard deviation computed from the data.

Clicking the **Data Analysis** button in the **Analysis** group under the **Data** tab opens the Data Analysis dialog box. Selecting the following tools in the **Analysis Tools** window and clicking **OK** opens their respective dialog boxes.

t-Test: Paired Two Sample for Means
returns P-values and critical t-values for one-tailed and two-tailed tests, the t value of the sample test statistic, and summary statistics for the data. This test is used for dependent samples.

t-Test: Two-Sample Assuming Equal Variances
returns P-values and critical t-values for one-tailed and two-tailed tests, the t value of the sample test statistic, and summary statistics for the data. This test is used for independent samples when the two standard deviations are unknown but assumed to be equal.

t-Test: Two-Sample Assuming Unequal Variances
returns P-values and critical t-values for one-tailed and two-tailed tests, the t value of the sample test statistic, and summary statistics for the data. This test is used for independent samples when the two standard deviations are unknown and assumed to be unequal.

z-Test: Two Sample for Means
returns P-values and critical z-values for one-tailed and two-tailed tests, the z value of the sample test statistic, and summary statistics for the data. This test is used for independent samples when the two standard deviations are known.

Copyright © Cengage Learning. All rights reserved.

Appendix: Descriptions of Data Sets in the Online Study Center

Appendix Preface

There are over 70 data sets saved in Excel, Minitab Portable, SPSS, TI-83 Plus, and TI-84 Plus/ASCII formats to accompany *Understanding Basic Statistics,* 5th edition. These files can be found on the Brase/Brase statistics site at www.cengage.com/statistics/Brase/UBS5e. The data sets are organized by category.

A. The following are provided for each data set:
    1. The category
    2. A brief description of the data and variables with a reference when appropriate
    3. File names for Excel, Minitab, SPSS, and TI-83 Plus and TI-84 Plus/ASCII formats

B. The categories are
    1. **Single variable**
        File name prefix **Sv** followed by the data set number
        41 data sets..page A-5
    2. **Two variable independent samples** (large and small samples)
        File name prefix **Tvis** followed by the data set number
        10 data sets..page A-19
    3. **Two variable dependent samples** appropriate for *t*-tests
        File name prefix **Tvds** followed by the data set number
        10 data sets..page A-24
    4. **Simple linear regression**
        File name prefix **Slr** followed by the data set number
        12 data sets..page A-28

C. The formats are
    1. Excel files in subdirectory Excel_5e. These files have suffix .xls
    2. Minitab portable files in subdirectory Minitab_5e. These files have suffix .mtp
    3. TI-83 Plus and TI-84 Plus/ASCII files in subdirectory TI8384_5e. These files have suffix .txt
    4. SPSS files in subdirectory SPSS_5e. These files have suffix .sav

Copyright © Cengage Learning. All rights reserved.

Suggestions for Using the Data Sets

1. Single variable (file name prefix Sv)

    These data sets are appropriate for:

    Graphs: Histograms, box plots

    Descriptive statistics: Mean, median, mode, variance, standard deviation, coefficient of variation, 5-number summary

    Inferential statistics: Confidence intervals for the population mean, hypothesis tests of a single mean

2. Two independent data sets (file name prefix Tvis)

    Graphs: Histograms, box plots for each data set

    Descriptive statistics: Mean, median, mode, variance, standard deviation, coefficient of variation, 5-number summary for each data set

    Inferential statistics: Confidence intervals for the difference of means, hypothesis tests for the difference of means

3. Paired data, dependent samples (file name prefix Tvds)

    Descriptive statistics: Mean, median, mode, variance, standard deviation, coefficient of variation, 5-number summary for the difference of the paired data values.

    Inferential statistics: Hypothesis tests for the difference of means (paired data)

4. Data pairs for simple linear regression (file name prefix Slr)

    Graphs: Scatter plots, for individual variables histograms and box plots

    Descriptive statistics:

    *   Mean, median, mode, variance, standard deviation, coefficient of variation, 5-number summary for individual variables.

    *   Least-squares line, sample correlation coefficient, sample coefficient of determination

Copyright © Cengage Learning. All rights reserved.

Descriptions of Data Sets

SINGLE VARIABLE

File name prefix: Sv followed by the number of the data file

01. **Disney Stock Volume (Single Variable)**
    The following data represent the number of shares of Disney stock (in hundreds of shares) sold for a random sample of 60 trading days
    Reference: *The Denver Post*, business section

12584	9441	18960	21480	10766	13059	8589	4965
4803	7240	10906	8561	6389	14372	18149	6309
13051	12754	10860	9574	19110	29585	21122	14522
17330	18119	10902	29158	16065	10376	10999	17950
15418	12618	16561	8022	9567	9045	8172	13708
11259	10518	9301	5197	11259	10518	9301	5197
6758	7304	7628	14265	13054	15336	14682	27804
16022	24009	32613	19111				

File names   Excel: Sv01.xls
Minitab: Sv01.mtp
SPSS: Sv01.sav
TI-83 Plus and TI-84 Plus/ASCII: Sv01.txt

02. **Weights of Pro Football Players (Single Variable)**
    The following data represent weights in pounds of 50 randomly selected pro football linebackers.
    Reference: The Sports Encyclopedia of Pro Football

225	230	235	238	232	227	244	222
250	226	242	253	251	225	229	247
239	223	233	222	243	237	230	240
255	230	245	240	235	252	245	231
235	234	248	242	238	240	240	240
235	244	247	250	236	246	243	255
241	245						

File names   Excel: Sv02.xls
Minitab: Sv02.mtp
SPSS: Sv02.sav
TI-83 Plus and TI-84 Plus/ASCII: Sv02.txt

Copyright © Cengage Learning. All rights reserved.

03. **Heights of Pro Basketball Players (Single Variable)**
The following data represent heights in feet of 65 randomly selected pro basketball players.
Reference: All-Time Player Directory, The Official NBA Encyclopedia

6.50	6.25	6.33	6.50	6.42	6.67	6.83	6.82
6.17	7.00	5.67	6.50	6.75	6.54	6.42	6.58
6.00	6.75	7.00	6.58	6.29	7.00	6.92	6.42
5.92	6.08	7.00	6.17	6.92	7.00	5.92	6.42
6.00	6.25	6.75	6.17	6.75	6.58	6.58	6.46
5.92	6.58	6.13	6.50	6.58	6.63	6.75	6.25
6.67	6.17	6.17	6.25	6.00	6.75	6.17	6.83
6.00	6.42	6.92	6.50	6.33	6.92	6.67	6.33
6.08							

File names   Excel: Sv03.xls
       Minitab: Sv03.mtp
       SPSS: Sv03.sav
       TI-83 Plus and TI-84 Plus/ASCII: Sv03.txt

04. **Miles per Gallon Gasoline Consumption (Single Variable)**
The following data represent miles per gallon gasoline consumption (highway) for a random sample of 55 makes and models of passenger cars.
Reference: Environmental Protection Agency

30	27	22	25	24	25	24	15
35	35	33	52	49	10	27	18
20	23	24	25	30	24	24	24
18	20	25	27	24	32	29	27
24	27	26	25	24	28	33	30
13	13	21	28	37	35	32	33
29	31	28	28	25	29	31	

File names   Excel: Sv04.xls
       Minitab: Sv04.mtp
       SPSS: Sv04.sav
       TI-83 Plus and TI-84 Plus/ASCII: Sv04.txt

05. **Fasting Glucose Blood Tests (Single Variable)**
The following data represent glucose blood levels (mg/100mL) after a 12-hour fast for a random sample of 70 women.
Reference: *American J. Clin. Nutr.*, Vol. 19, pp. 345-351

45	66	83	71	76	64	59	59
76	82	80	81	85	77	82	90
87	72	79	69	83	71	87	69
81	76	96	83	67	94	101	94
89	94	73	99	93	85	83	80
78	80	85	83	84	74	81	70
65	89	70	80	84	77	65	46
80	70	75	45	101	71	109	73
73	80	72	81	63	74		

File names   Excel: Sv05.xls
       Minitab: Sv05.mtp
       SPSS: Sv05.sav
       TI-83 Plus and TI-84 Plus/ASCII: Sv05.txt

Copyright © Cengage Learning. All rights reserved.

06. Number of Children in Rural Canadian Families (Single Variable)

The following data represent the numbers of children in a random sample of 50 rural Canadian families.

Reference: *American Journal of Sociology*, Vol. 53, pp. 470-480

11	13	4	14	10	2	5	0
0	3	9	2	5	2	3	3
3	4	7	1	9	4	3	3
2	6	0	2	6	5	9	5
4	3	2	5	2	2	3	5
14	7	6	6	2	5	3	4
6	1						

File names   Excel: Sv06.xls
Minitab: Sv06.mtp
SPSS: Sv06.sav
TI-83 Plus and TI-84 Plus/ASCII: Sv06.txt

07. Children as a % of Population (Single Variable)

The following data represent percentages of children in the population for a random sample of 72 Denver neighborhoods.

Reference: The Piton Foundation, Denver, Colorado

30.2	18.6	13.6	36.9	32.8	19.4	12.3	39.7	22.2	31.2
36.4	37.7	38.8	28.1	18.3	22.4	26.5	20.4	37.6	23.8
22.1	53.2	6.8	20.7	31.7	10.4	21.3	19.6	41.5	29.8
14.7	12.3	17.0	16.7	20.7	34.8	7.5	19.0	27.2	16.3
24.3	39.8	31.1	34.3	15.9	24.2	20.3	31.2	30.0	33.1
29.1	39.0	36.0	31.8	32.9	26.5	4.9	19.5	21.0	24.2
12.1	38.3	39.3	20.2	24.0	28.6	27.1	30.0	60.8	39.2
21.6	20.3								

File names   Excel: Sv07.xls
Minitab: Sv07.mtp
SPSS: Sv07.sav
TI-83 Plus and TI-84 Plus/ASCII: Sv07.txt

08. Percentage Change in Household Income (Single Variable)

The following data represent the percentage change in household income over a five-year period for a random sample of $n = 78$ Denver neighborhoods.

Reference: The Piton Foundation, Denver, Colorado

27.2	25.2	25.7	80.9	26.9	20.2	25.4	26.9	26.4	26.3
27.5	38.2	20.9	31.3	23.5	26.0	35.8	30.9	15.5	24.8
29.4	11.7	32.6	32.2	27.6	27.5	28.7	28.0	15.6	20.0
21.8	18.4	27.3	13.4	14.7	21.6	26.8	20.9	32.7	29.3
21.4	29.0	7.2	25.7	25.5	39.8	26.6	24.2	33.5	16.0
29.4	26.8	32.0	24.7	24.2	29.8	25.8	18.2	26.0	26.2
21.7	27.0	23.7	28.0	11.2	26.2	21.6	23.7	28.3	34.1
40.8	16.0	50.5	54.1	3.3	23.5	10.1	14.8		

File names   Excel: Sv08.xls
Minitab: Sv08.mtp
SPSS: Sv08.sav
TI-83 Plus and TI-84 Plus/ASCII: Sv08.txt

Copyright © Cengage Learning. All rights reserved.

09. **Crime Rate per 1,000 Population (Single Variable)**
    The following data represent the crime rates per 1,000 population for a random sample of 70 Denver neighborhoods.
    Reference: The Piton Foundation, Denver, Colorado

84.9	45.1	132.1	104.7	258.0	36.3	26.2	207.7
58.5	65.3	42.5	53.2	172.6	69.2	179.9	65.1
32.0	38.3	185.9	42.4	63.0	86.4	160.4	26.9
154.2	111.0	139.9	68.2	127.0	54.0	42.1	105.2
77.1	278.0	73.0	32.1	92.7	704.1	781.8	52.2
65.0	38.6	22.5	157.3	63.1	289.1	52.7	108.7
66.3	69.9	108.7	96.9	27.1	105.1	56.2	80.1
59.6	77.5	68.9	35.2	65.4	123.2	130.8	70.7
25.1	62.6	68.6	334.5	44.6	87.1		

File names   Excel: Sv09.xls
                    Minitab: Sv09.mtp
                    SPSS: Sv09.sav
                    TI-83 Plus and TI-84 Plus/ASCII: Sv09.txt

10. **Percentage Change in Population (Single Variable)**
    The following data represent the percentage change in population over a nine-year period for a random sample of 64 Denver neighborhoods.
    Reference: The Piton Foundation, Denver, Colorado

6.2	5.4	8.5	1.2	5.6	28.9	6.3	10.5	-1.5	17.3
21.6	-2.0	-1.0	3.3	2.8	3.3	28.5	-0.7	8.1	32.6
68.6	56.0	19.8	7.0	38.3	41.2	4.9	7.8	7.8	97.8
5.5	21.6	32.5	-0.5	2.8	4.9	8.7	-1.3	4.0	32.2
2.0	6.4	7.1	8.8	3.0	5.1	-1.9	-2.6	1.6	7.4
10.8	4.8	1.4	19.2	2.7	71.4	2.5	6.2	2.3	10.2
1.9	2.3	-3.3	2.6						

File names   Excel: Sv10.xls
                    Minitab: Sv10.mtp
                    SPSS: Sv10.sav
                    TI-83 Plus and TI-84 Plus/ASCII: Sv10.txt

11. **Thickness of the Ozone Column (Single Variable)**
    The following data represent the January mean thickness of the ozone column above Arosa, Switzerland (Dobson units: one milli-centimeter ozone at standard temperature and pressure). The data is from a random sample of years from 1926 on.
    Reference: Laboratorium für Atmosphärensphysik, Switzerland

324	332	362	383	335	349	354	319	360	329
400	341	315	368	361	336	349	347	338	332
341	352	342	361	318	337	300	352	340	371
327	357	320	377	338	361	301	331	334	387
336	378	369	332	344					

File names   Excel: Sv11.xls
                    Minitab: Sv11.mtp
                    SPSS: Sv11.sav
                    TI-83 Plus and TI-84 Plus/ASCII: Sv11.txt

Copyright © Cengage Learning. All rights reserved.

12. **Sun Spots (Single Variable)**

The following data represent the January mean number of sunspots. The data are taken from a random sample of Januarys from 1749 to 1983.

Reference: Waldmeir, M., *Sun Spot Activity*, International Astronomical Union Bulletin

12.5	14.1	37.6	48.3	67.3	70.0	43.8	56.5	59.7	24.0
12.0	27.4	53.5	73.9	104.0	54.6	4.4	177.3	70.1	54.0
28.0	13.0	6.5	134.7	114.0	72.7	81.2	24.1	20.4	13.3
9.4	25.7	47.8	50.0	45.3	61.0	39.0	12.0	7.2	11.3
22.2	26.3	34.9	21.5	12.8	17.7	34.6	43.0	52.2	47.5
30.9	11.3	4.9	88.6	188.0	35.6	50.5	12.4	3.7	18.5
115.5	108.5	119.1	101.6	59.9	40.7	26.5	23.1	73.6	165.0
202.5	217.4	57.9	38.7	15.3	8.1	16.4	84.3	51.9	58.0
74.7	96.0	48.1	51.1	31.5	11.8	4.5	78.1	81.6	68.9

File names   Excel: Sv12.xls
                    Minitab: Sv12.mtp
                    SPSS: Sv12.sav
                    TI-83 Plus and TI-84 Plus/ASCII: Sv12.txt

13. **Motion of Stars (Single Variable)**

The following data represent the angular motions of stars across the sky due to each star's own velocity. A random sample of stars from the M92 globular cluster was used. Units are arc seconds per century.

Reference: Cudworth, K.M., *Astronomical Journal*, Vol. 81, pp. 975-982

0.042	0.048	0.019	0.025	0.028	0.041	0.030	0.051	0.026
0.040	0.018	0.022	0.048	0.045	0.019	0.028	0.029	0.018
0.033	0.035	0.019	0.046	0.021	0.026	0.026	0.033	0.046
0.023	0.036	0.024	0.014	0.012	0.037	0.034	0.032	0.035
0.015	0.027	0.017	0.035	0.021	0.016	0.036	0.029	0.031
0.016	0.024	0.015	0.019	0.037	0.016	0.024	0.029	0.025
0.022	0.028	0.023	0.021	0.020	0.020	0.016	0.016	0.016
0.040	0.029	0.025	0.025	0.042	0.022	0.037	0.024	0.046
0.016	0.024	0.028	0.027	0.060	0.045	0.037	0.027	0.028
0.022	0.048	0.053						

File names   Excel: Sv13.xls
                    Minitab: Sv13.mtp
                    SPSS: Sv13.sav
                    TI-83 Plus and TI-84 Plus/ASCII: Sv13.txt

Copyright © Cengage Learning. All rights reserved.

14. **Arsenic and Ground Water (Single Variable)**
   The following data represent (naturally occurring) concentrations of arsenic in ground water for a random sample of 102 Northwest Texas wells. Units are parts per billion.
   Reference: Nichols, C.E. and Kane, V.E., Union Carbide Technical Report K/UR-1

7.6	10.4	13.5	4.0	19.9	16.0	12.0	12.2	11.4	12.7
3.0	10.3	21.4	19.4	9.0	6.5	10.1	8.7	9.7	6.4
9.7	63.0	15.5	10.7	18.2	7.5	6.1	6.7	6.9	0.8
73.5	12.0	28.0	12.6	9.4	6.2	15.3	7.3	10.7	15.9
5.8	1.0	8.6	1.3	13.7	2.8	2.4	1.4	2.9	13.1
15.3	9.2	11.7	4.5	1.0	1.2	0.8	1.0	2.4	4.4
2.2	2.9	3.6	2.5	1.8	5.9	2.8	1.7	4.6	5.4
3.0	3.1	1.3	2.6	1.4	2.3	1.0	5.4	1.8	2.6
3.4	1.4	10.7	18.2	7.7	6.5	12.2	10.1	6.4	10.7
6.1	0.8	12.0	28.1	9.4	6.2	7.3	9.7	62.1	15.5
6.4	9.5								

   File names   Excel: Sv14.xls
                     Minitab: Sv14.mtp
                     SPSS: Sv14.sav
                     TI-83 Plus and TI-84 Plus/ASCII: Sv14.txt

15. **Uranium in Ground Water (Single Variable)**
   The following data represent (naturally occurring) concentrations of uranium in ground water for a random sample of 100 Northwest Texas wells. Units are parts per billion.
   Reference: Nichols, C.E. and Kane, V.E., Union Carbide Technical Report K/UR-1

8.0	13.7	4.9	3.1	78.0	9.7	6.9	21.7	26.8
56.2	25.3	4.4	29.8	22.3	9.5	13.5	47.8	29.8
13.4	21.0	26.7	52.5	6.5	15.8	21.2	13.2	12.3
5.7	11.1	16.1	11.4	18.0	15.5	35.3	9.5	2.1
10.4	5.3	11.2	0.9	7.8	6.7	21.9	20.3	16.7
2.9	124.2	58.3	83.4	8.9	18.1	11.9	6.7	9.8
15.1	70.4	21.3	58.2	25.0	5.5	14.0	6.0	11.9
15.3	7.0	13.6	16.4	35.9	19.4	19.8	6.3	2.3
1.9	6.0	1.5	4.1	34.0	17.6	18.6	8.0	7.9
56.9	53.7	8.3	33.5	38.2	2.8	4.2	18.7	12.7
3.8	8.8	2.3	7.2	9.8	7.7	27.4	7.9	11.1
24.7								

   File names   Excel: Sv15.xls
                     Minitab: Sv15.mtp
                     SPSS: Sv15.sav
                     TI-83 Plus and TI-84 Plus/ASCII: Sv15.txt

Copyright © Cengage Learning. All rights reserved.

16. **Ground Water pH (Single Variable)**

    A pH less than 7 is acidic, and a pH above 7 is alkaline. The following data represent pH levels in ground water for a random sample of 102 Northwest Texas wells.

    Reference: Nichols, C.E. and Kane, V.E., Union Carbide Technical Report K/UR-1

7.6	7.7	7.4	7.7	7.1	8.2	7.4	7.5	7.2	7.4
7.2	7.6	7.4	7.8	8.1	7.5	7.1	8.1	7.3	8.2
7.6	7.0	7.3	7.4	7.8	8.1	7.3	8.0	7.2	8.5
7.1	8.2	8.1	7.9	7.2	7.1	7.0	7.5	7.2	7.3
8.6	7.7	7.5	7.8	7.6	7.1	7.8	7.3	8.4	7.5
7.1	7.4	7.2	7.4	7.3	7.7	7.0	7.3	7.6	7.2
8.1	8.2	7.4	7.6	7.3	7.1	7.0	7.0	7.4	7.2
8.2	8.1	7.9	8.1	8.2	7.7	7.5	7.3	7.9	8.8
7.1	7.5	7.9	7.5	7.6	7.7	8.2	8.7	7.9	7.0
8.8	7.1	7.2	7.3	7.6	7.1	7.0	7.0	7.3	7.2
7.8	7.6								

    File names   Excel: Sv16.xls
                        Minitab: Sv16.mtp
                        SPSS: Sv16.sav
                        TI-83 Plus and TI-84 Plus/ASCII: Sv16.txt

17. **Static Fatigue 90% Stress Level (Single Variable)**

    Kevlar Epoxy is a material used on the NASA space shuttle. Strands of this epoxy were tested at 90% breaking strength. The following data represent time to failure in hours at the 90% stress level for a random sample of 50 epoxy strands.

    Reference: R.E. Barlow, University of California, Berkeley

0.54	1.80	1.52	2.05	1.03	1.18	0.80	1.33	1.29	1.11
3.34	1.54	0.08	0.12	0.60	0.72	0.92	1.05	1.43	3.03
1.81	2.17	0.63	0.56	0.03	0.09	0.18	0.34	1.51	1.45
1.52	0.19	1.55	0.02	0.07	0.65	0.40	0.24	1.51	1.45
1.60	1.80	4.69	0.08	7.89	1.58	1.64	0.03	0.23	0.72

    File names   Excel: Sv17.xls
                        Minitab: Sv17.mtp
                        SPSS: Sv17.sav
                        TI-83 Plus and TI-84 Plus/ASCII: Sv17.txt

18. **Static Fatigue 80% Stress Level (Single Variable)**

    Kevlar Epoxy is a material used on the NASA space shuttle. Strands of this epoxy were tested at 80% breaking strength. The following data represent time to failure in hours at the 80% stress level for a random sample of 54 epoxy strands.

    Reference: R.E. Barlow, University of California, Berkeley

152.2	166.9	183.8	8.5	1.8	118.0	125.4	132.8	10.6
29.6	50.1	202.6	177.7	160.0	87.1	112.6	122.3	124.4
131.6	140.9	7.5	41.9	59.7	80.5	83.5	149.2	137.0
301.1	329.8	461.5	739.7	304.3	894.7	220.2	251.0	269.2
130.4	77.8	64.4	381.3	329.8	451.3	346.2	663.0	49.1
31.7	116.8	140.2	334.1	285.9	59.7	44.1	351.2	93.2

    File names   Excel: Sv18.xls
                        Minitab: Sv18.mtp
                        SPSS: Sv18.sav
                        TI-83 Plus and TI-84 Plus/ASCII: Sv18.txt

Copyright © Cengage Learning. All rights reserved.

19. **Tumor Recurrence (Single Variable)**
    Certain kinds of tumors tend to recur. The following data represent the lengths of time in months for a tumor to recur after chemotherapy (sample size: 42).
    Reference: Byar, D.P., *Urology,* Vol. 10, pp. 556-561

19	18	17	1	21	22	54	46	25	49
50	1	59	39	43	39	5	9	38	18
14	45	54	59	46	50	29	12	19	36
38	40	43	41	10	50	41	25	19	39
27	20								

    File names   Excel: Sv19.xls
                    Minitab: Sv19.mtp
                    SPSS: Sv19.sav
                    TI-83 Plus and TI-84 Plus/ASCII: Sv19.txt

20. **Weight of Harvest (Single Variable)**
    The following data represent the weights in kilograms of maize harvest from a random sample of 72 experimental plots on the island of St. Vincent (Caribbean).
    Reference: Springer, B.G.F., *Proceedings, Caribbean Food Corps. Soc.,* Vol. 10, pp. 147-152

24.0	27.1	26.5	13.5	19.0	26.1	23.8	22.5	20.0
23.1	23.8	24.1	21.4	26.7	22.5	22.8	25.2	20.9
23.1	24.9	26.4	12.2	21.8	19.3	18.2	14.4	22.4
16.0	17.2	20.3	23.8	24.5	13.7	11.1	20.5	19.1
20.2	24.1	10.5	13.7	16.0	7.8	12.2	12.5	14.0
22.0	16.5	23.8	13.1	11.5	9.5	22.8	21.1	22.0
11.8	16.1	10.0	9.1	15.2	14.5	10.2	11.7	14.6
15.5	23.7	25.1	29.5	24.5	23.2	25.5	19.8	17.8

    File names   Excel: Sv20.xls
                    Minitab: Sv20.mtp
                    SPSS: Sv20.sav
                    TI-83 Plus and TI-84 Plus/ASCII: Sv20.txt

21. **Apple Trees (Single Variable)**
    The following data represent the trunk girths (mm) from a random sample of 60 four-year-old apple trees at East Malling Research Station (England)
    Reference: S.C. Pearce, University of Kent at Canterbury

108	99	106	102	115	120	120	117	122	142
106	111	119	109	125	108	116	105	117	123
103	114	101	99	112	120	108	91	115	109
114	105	99	122	106	113	114	75	96	124
91	102	108	110	83	90	69	117	84	142
122	113	105	112	117	122	129	100	138	117

    File names   Excel: Sv21.xls
                    Minitab: Sv21.mtp
                    SPSS: Sv21.sav
                    TI-83 Plus and TI-84 Plus/ASCII: Sv21.txt

Copyright © Cengage Learning. All rights reserved.

22. **Black Mesa Archaeology (Single Variable)**
    The following data represent rim diameters (cm) from a random sample of 40 bowls found at Black Mesa archaeological site. The diameters are estimated from broken pot shards.
    Reference: Michelle Hegmon, Crow Canyon Archaeological Center, Cortez, Colorado

17.2	15.1	13.8	18.3	17.5	11.1	7.3	23.1	21.5	19.7
17.6	15.9	16.3	25.7	27.2	33.0	10.9	23.8	24.7	18.6
16.9	18.8	19.2	14.6	8.2	9.7	11.8	13.3	14.7	15.8
17.4	17.1	21.3	15.2	16.8	17.0	17.9	18.3	14.9	17.7

    File names   Excel: Sv22.xls
                        Minitab: Sv22.mtp
                        SPSS: Sv22.sav
                        TI-83 Plus and TI-84 Plus/ASCII: Sv22.txt

23. **Wind Mountain Archaeology (Single Variable)**
    The following data represent depths (cm) for a random sample of 73 significant archaeological artifacts at the Wind Mountain excavation site.
    Reference: Woosley, A. and McIntyre, A., *Mimbres Mogolion Archaeology*, University of New Mexico Press

85	45	75	60	90	90	115	30	55	58
78	120	80	65	65	140	65	50	30	125
75	137	80	120	15	45	70	65	50	45
95	70	70	28	40	125	105	75	80	70
90	68	73	75	55	70	95	65	200	75
15	90	46	33	100	65	60	55	85	50
10	68	99	145	45	75	45	95	85	65
65	52	82							

    File names   Excel: Sv23.xls
                        Minitab: Sv23.mtp
                        SPSS: Sv23.sav
                        TI-83 Plus and TI-84 Plus/ASCII: Sv23.txt

24. **Arrow Heads (Single Variable)**
    The following data represent the lengths (cm) of a random sample of 61 projectile points found at the Wind Mountain Archaeological site.
    Reference: Woosley, A. and McIntyre, A., *Mimbres Mogolion Archaeology*, University of New Mexico Press

3.1	4.1	1.8	2.1	2.2	1.3	1.7	3.0	3.7	2.3
2.6	2.2	2.8	3.0	3.2	3.3	2.4	2.8	2.8	2.9
2.9	2.2	2.4	2.1	3.4	3.1	1.6	3.1	3.5	2.3
3.1	2.7	2.1	2.0	4.8	1.9	3.9	2.0	5.2	2.2
2.6	1.9	4.0	3.0	3.4	4.2	2.4	3.5	3.1	3.7
3.7	2.9	2.6	3.6	3.9	3.5	1.9	4.0	4.0	4.6
1.9									

    File names   Excel: Sv24.xls
                        Minitab: Sv24.mtp
                        SPSS: Sv24.sav
                        TI-83 Plus and TI-84 Plus/ASCII: Sv24.txt

Copyright © Cengage Learning. All rights reserved.

25. **Anasazi Indian Bracelets (Single Variable)**
    The following data represent the diameters (cm) of shell bracelets and rings found at the Wind Mountain archaeological site.
    Reference: Woosley, A. and McIntyre, A., *Mimbres Mogolion Archaeology*, University of New Mexico Press

5.0	5.0	8.0	6.1	6.0	5.1	5.9	6.8	4.3	5.5
7.2	7.0	5.0	5.6	5.3	7.0	3.4	8.2	4.3	5.2
1.5	6.1	4.0	6.0	5.5	5.2	5.2	5.2	5.5	7.2
6.0	6.2	5.2	5.0	4.0	5.7	5.1	6.1	5.7	7.3
7.3	6.7	4.2	4.0	6.0	7.1	7.3	5.5	5.8	8.9
7.5	8.3	6.8	4.9	4.0	6.2	7.7	5.0	5.2	6.8
6.1	7.2	4.4	4.0	5.0	6.0	6.2	7.2	5.8	6.8
7.7	4.7	5.3							

    File names   Excel: Sv25.xls
                        Minitab: Sv25.mtp
                        SPSS: Sv25.sav
                        TI-83 Plus and TI-84 Plus/ASCII: Sv25.txt

26. **Pizza Franchise Fees (Single Variable)**
    The following data represent annual franchise fees (in thousands of dollars) for a random sample of 36 pizza franchises.
    Reference: *Business Opportunities Handbook*

25.0	15.5	7.5	19.9	18.5	25.5	15.0	5.5	15.2	15.0
14.9	18.5	14.5	29.0	22.5	10.0	25.0	35.5	22.1	89.0
17.5	33.3	17.5	12.0	15.5	25.5	12.5	17.5	12.5	35.0
30.0	21.0	35.5	10.5	5.5	20.0				

    File names   Excel: Sv26.xls
                        Minitab: Sv26.mtp
                        SPSS: Sv26.sav
                        TI-83 Plus and TI-84 Plus/ASCII: Sv26.txt

27. **Pizza Franchise Start-up Requirement (Single Variable)**
    The following data represent the start-up costs (in thousands of dollars) for a random sample of 36 pizza franchises.
    Reference: *Business Opportunities Handbook*

40	25	50	129	250	128	110	142	25	90
75	100	500	214	275	50	128	250	50	75
30	40	185	50	175	125	200	150	150	120
95	30	400	149	235	100				

    File names   Excel: Sv27.xls
                        Minitab: Sv27.mtp
                        SPSS: Sv27.sav
                        TI-83 Plus and TI-84 Plus/ASCII: Sv27.txt

Copyright © Cengage Learning. All rights reserved.

28. College Degrees (Single Variable)

The following data represent percentages of the adult population with college degrees. The sample is from a random collection of 68 Midwest counties.

Reference: *County and City Data Book,* 12th edition, U.S. Department of Commerce

9.9	9.8	6.8	8.9	11.2	15.5	9.8	16.8	9.9	11.6
9.2	8.4	11.3	11.5	15.2	10.8	16.3	17.0	12.8	11.0
6.0	16.0	12.1	9.8	9.4	9.9	10.5	11.8	10.3	11.1
12.5	7.8	10.7	9.6	11.6	8.8	12.3	12.2	12.4	10.0
10.0	18.1	8.8	17.3	11.3	14.5	11.0	12.3	9.1	12.7
5.6	11.7	16.9	13.7	12.5	9.0	12.7	11.3	19.5	30.7
9.4	9.8	15.1	12.8	12.9	17.5	12.3	8.2		

File names   Excel: Sv28.xls
                               Minitab: Sv28.mtp
                               SPSS: Sv28.sav
                               TI-83 Plus and TI-84 Plus/ASCII: Sv28.txt

29. Poverty Level (Single Variable)

The following data represent percentages of all persons below the poverty level. The sample is from a random collection of 80 cities in the Western U.S.

Reference: *County and City Data Book,* 12th edition, U.S. Department of Commerce

12.1	27.3	20.9	14.9	4.4	21.8	7.1	16.4	13.1
9.4	9.8	15.7	29.9	8.8	32.7	5.1	9.0	16.8
21.6	4.2	11.1	14.1	30.6	15.4	20.7	37.3	7.7
19.4	18.5	19.5	8.0	7.0	20.2	6.3	12.9	13.3
30.0	4.9	14.4	14.1	22.6	18.9	16.8	11.5	19.2
21.0	11.4	7.8	6.0	37.3	44.5	37.1	28.7	9.0
17.9	16.0	20.2	11.5	10.5	17.0	3.4	3.3	15.6
16.6	29.6	14.9	23.9	13.6	7.8	14.5	19.6	31.5
28.1	19.2	4.9	12.7	15.1	9.6	23.8	10.1	

File names   Excel: Sv29.xls
                               Minitab: Sv29.mtp
                               SPSS: Sv29.sav
                               TI-83 Plus and TI-84 Plus/ASCII: Sv29.txt

30. Working at Home (Single Variable)

The following data represent percentages of adults whose primary employment involves working at home. The data are from a random sample of 50 California cities.

Reference: *County and City Data Book,* 12th edition, U.S. Department of Commerce

4.3	5.1	3.1	8.7	4.0	5.2	11.8	3.4	8.5	3.0
4.3	6.0	3.7	3.7	4.0	3.3	2.8	2.8	2.6	4.4
7.0	8.0	3.7	3.3	3.7	4.9	3.0	4.2	5.4	6.6
2.4	2.5	3.5	3.3	5.5	9.6	2.7	5.0	4.8	4.1
3.8	4.8	14.3	9.2	3.8	3.6	6.5	2.6	3.5	8.6

File names   Excel: Sv30.xls
                               Minitab: Sv30.mtp
                               SPSS: Sv30.sav
                               TI-83 Plus and TI-84 Plus/ASCII: Sv30.txt

Copyright © Cengage Learning. All rights reserved.

31. **Number of Pups in Wolf Den (Single Variable)**
    The following data represent numbers of wolf pups per den from a random sample of 16 wolf dens.
    Reference: *The Wolf in the Southwest: The Making of an Endangered Species*, Brown, D.E., University of Arizona Press

5	8	7	5	3	4	3	9
5	8	5	6	5	6	4	7

    File names   Excel: Sv31.xls
            Minitab: Sv31.mtp
            SPSS: Sv31.sav
            TI-83 Plus and TI-84 Plus/ASCII: Sv31.txt

32. **Glucose Blood Level (Single Variable)**
    The following data represent glucose blood levels (mg/100ml) after a 12-hour fast for a random sample of 6 tests given to an individual adult female.
    Reference: *American J. Clin. Nutr., Vol. 19*, pp. 345-351

83	83	86	86	78	88

    File names   Excel: Sv32.xls
            Minitab: Sv32.mtp
            SPSS: Sv32.sav
            TI-83 Plus and TI-84 Plus/ASCII: Sv32.txt

33. **Length of Remission (Single Variable)**
    The drug 6-mP (6-mercaptopurine) is used to treat leukemia. The following data represent the lengths of remission in weeks for a random sample of 21 patients using 6-mP.
    Reference: E.A. Gehan, University of Texas Cancer Center

10	7	32	23	22	6	16	34	32	25
11	20	19	6	17	35	6	13	9	6
10

    File names   Excel: Sv33.xls
            Minitab: Sv33.mtp
            SPSS: Sv33.sav
            TI-83 Plus and TI-84 Plus/ASCII: Sv33.txt

34. **Entry Level Jobs (Single Variable)**
    The following data represent percentages of entry-level jobs in a random sample of 16 Denver neighborhoods.
    Reference: The Piton Foundation, Denver, Colorado

8.9	22.6	18.5	9.2	8.2	24.3	15.3	3.7
9.2	14.9	4.7	11.6	16.5	11.6	9.7	8.0

    File names   Excel: Sv34.xls
            Minitab: Sv34.mtp
            SPSS: Sv34.sav
            TI-83 Plus and TI-84 Plus/ASCII: Sv34.txt

Copyright © Cengage Learning. All rights reserved.

35. Licensed Child Care Slots (Single Variable)
The following data represent the number of licensed childcare slots in a random sample of 15 Denver neighborhoods.
Reference: The Piton Foundation, Denver, Colorado

```
523  106  184  121  357  319  656  170
241  226  741  172  266  423  212
```

File names   Excel: Sv35.xls
                    Minitab: Sv35.mtp
                    SPSS: Sv35.sav
                    TI-83 Plus and TI-84 Plus/ASCII: Sv35.txt

36. Subsidized Housing (Single Variable)
The following data represent the percentages of subsidized housing in a random sample of 14 Denver neighborhoods.
Reference: The Piton Foundation, Denver, Colorado

```
10.2  11.8   9.7  22.3   6.8  10.4  11.0
 5.4   6.6  13.7  13.6   6.5  16.0  24.8
```

File names   Excel: Sv36.xls
                    Minitab: Sv36.mtp
                    SPSS: Sv36.sav
                    TI-83 Plus and TI-84 Plus/ASCII: Sv36.txt

37. Sulfate in Ground Water (Single Variable)
The following data represent naturally occurring amounts of sulfate (SO_4) in well water. Units: parts per million. The data are from a random sample of 24 water wells in Northwest Texas.
Reference: Nichols, C.E. and Kane, V.E., Union Carbide Corporation Technical Report K/UR-1

```
1850  1150  1340  1325  2500  1060  1220  2325  460
2000  1500  1775   620  1950   780   840  2650  975
 860   495  1900  1220  2125   990
```

File names   Excel: Sv37.xls
                    Minitab: Sv37.mtp
                    SPSS: Sv37.sav
                    TI-83 Plus and TI-84 Plus/ASCII: Sv37.txt

38. Earth's Rotation Rate (Single Variable)
The following data represent changes in the earth's rotation (i.e., day length). Units: 0.00001 second. The data are from a random sample of 23 years.
Reference: *Acta Astron. Sinica*, Vol. 15, pp. 79-85

```
-12  110   78  126  -35  104  111   22  -31   92
 51   36  231  -13   65  119   21  104  112  -15
137  139  101
```

File names   Excel: Sv38.xls
                    Minitab: Sv38.mtp
                    SPSS: Sv38.sav
                    TI-83 Plus and TI-84 Plus/ASCII: Sv38.txt

Copyright © Cengage Learning. All rights reserved.

39. Blood Glucose (Single Variable)

The following data represent glucose levels (mg/100ml) in the blood for a random sample of 27 non-obese adult subjects.

Reference: *Diabetologia*, Vol. 16, pp. 17-24

80	85	75	90	70	97	91	85	90	85
105	86	78	92	93	90	80	102	90	90
99	93	91	86	98	86	92			

File names   Excel: Sv39.xls
Minitab: Sv39.mtp
SPSS: Sv39.sav
TI-83 Plus and TI-84 Plus/ASCII: Sv39.txt

40. Plant Species (Single Variable)

The following data represent the observed number of native plant species from random samples of study plots on different islands in the Galapagos Island chain.

Reference: *Science*, Vol. 179, pp. 893-895

23	26	33	73	21	35	30	16	3	17
9	8	9	19	65	12	11	89	81	7
23	95	4	37	28					

File names   Excel: Sv40.xls
Minitab: Sv40.mtp
SPSS: Sv40.sav
TI-83 Plus and TI-84 Plus/ASCII: Sv40.txt

41. Apples (Single Variable)

The following data represent mean fruit weights (grams) of apples per tree for a random sample of 28 trees in an agricultural experiment.

Reference: *Aust. J. Agric Res.*, Vol. 25, p783-790

85.3	86.9	96.8	108.5	113.8	87.7	94.5	99.9	92.9
67.3	90.6	129.8	48.9	117.5	100.8	94.5	94.4	98.9
96.0	99.4	79.1	108.5	84.6	117.5	70.0	104.4	127.1
135.0								

File names   Excel: Sv41.xls
Minitab: Sv41.mtp
SPSS: Sv41.sav
TI-83 Plus and TI-84 Plus/ASCII: Sv41.txt

Copyright © Cengage Learning. All rights reserved.

TWO VARIABLE

ZZ	SINGLE	S

the data file

football players and 40

tball Encyclopedia

Customer:

Emily J Wild

Technology Guide Excel for Brase/Brase's Understanding Basic Statistics, Brief, 5th

Charles Henry Brase, Corrinne Pellillo Brase

E3-S058-F4

0D-DEH-606

No CD

Used - Good

9780547189161

Picker Notes.

M _____ 2 _____

WT _____ 2 _____

CC _____

iris virginica and a

38127397

[Amazon] betterworldbooks_: 103-3967246-0513840

1 Item

1034724072

QQ Late 530pm TUES, ZZ Late
230am WED

Ship. Created: 8/2/2015 1:31:00 PM

Copyright © Cengage Learning. All rights reserved.

03.   **Sepal Width Of *Iris Versicolor* Versus *Iris Virginica***
       (Two variable independent large samples)
       The following data represent sepal widths (cm) for a random sample of 40 *iris versicolor* and a random sample of 42 *iris virginica*.
       Reference: Anderson, E., *Bull. Amer. Iris Soc.*

       X1 = sepal width (cm.) *iris versicolor*
       3.2 3.2 3.1 2.3 2.8 2.8 3.3 2.4 2.9 2.7 2.0 3.0 2.2 2.9 2.9 3.1
       3.0 2.7 2.2 2.5 3.2 2.8 2.5 2.8 2.9 3.0 2.8 3.0 2.9 2.6 2.4 2.4
       2.7 2.7 3.0 3.4 3.1 2.3 3.0 2.5

       X2 = sepal width (cm.) *iris virginica*
       3.3 2.7 3.0 2.9 3.0 3.0 2.5 2.9 2.5 3.6 3.2 2.7 3.0 2.5 2.8 3.2
       3.0 3.8 2.6 2.2 3.2 2.8 2.8 2.7 3.3 3.2 2.8 3.0 2.8 3.0 2.8 3.8
       2.8 2.8 2.6 3.0 3.4 3.1 3.0 3.1 3.1 3.1

       File names   Excel: Tvis03.xls
                           Minitab: Tvis03.mtp
                           SPSS: Tvis03.sav
                           TI-83 Plus and TI-84 Plus/ASCII:
                                   X1 data is stored in Tvis03L1.txt
                                   X2 data is stored in Tvis03L2.txt

04.   **Archaeology, Ceramics (Two variable independent large samples)**
       The following data represent independent random samples of shard counts of painted ceramics found at the Wind Mountain archaeological site.
       Reference: Woosley, A. and McIntyre, A., *Mimbres Mogollon Archaeology*, University of New Mexico Press

       X1 = count Mogollon red on brown
       52   10   8   71   7   31   24   20   17   5
       16   75   25   17   14   33   13   17   12   19
       67   13   35   14   3   7   9   19   16   22
        7   10   9   49   6   13   24   45   14   20
        3   6   30   41   26   32   14   33   1   48
       44   14   16   15   13   8   61   11   12   16
       20   39

       X2 = count Mimbres black on white
       61   21   78   9   14   12   34   54   10   15
       43   9   7   67   18   18   24   54   8   10
       16   6   17   14   25   22   25   13   23   12
       36   10   56   35   79   69   41   36   18   25
       27   27   11   13

       File names   Excel: Tvis04.xls
                           Minitab: Tvis04.mtp
                           SPSS: Tvis04.sav
                           TI-83 Plus and TI-84 Plus/ASCII:
                                   X1 data is stored in Tvis04L1.txt
                                   X2 data is stored in Tvis04L2.txt

Copyright © Cengage Learning. All rights reserved.

05. **Agriculture, Water Content of Soil (Two variable independent large samples)**
The following data represent soil water content (% water by volume) for independent random samples of soil from two experimental fields growing bell peppers.
Reference: *Journal of Agricultural, Biological, and Environmental Statistics*, Vol. 2, No. 2, pp. 149-155

X1 = soil water content from field I

15.1	11.2	10.3	10.8	16.6	8.3	9.1	12.3	9.1	14.3
10.7	16.1	10.2	15.2	8.9	9.5	9.6	11.3	14.0	11.3
15.6	11.2	13.8	9.0	8.4	8.2	12.0	13.9	11.6	16.0
9.6	11.4	8.4	8.0	14.1	10.9	13.2	13.8	14.6	10.2
11.5	13.1	14.7	12.5	10.2	11.8	11.0	12.7	10.3	10.8
11.0	12.6	10.8	9.6	11.5	10.6	11.7	10.1	9.7	9.7
11.2	9.8	10.3	11.9	9.7	11.3	10.4	12.0	11.0	10.7
8.8	11.1								

X2 = soil water content from field II

12.1	10.2	13.6	8.1	13.5	7.8	11.8	7.7	8.1	9.2
14.1	8.9	13.9	7.5	12.6	7.3	14.9	12.2	7.6	8.9
13.9	8.4	13.4	7.1	12.4	7.6	9.9	26.0	7.3	7.4
14.3	8.4	13.2	7.3	11.3	7.5	9.7	12.3	6.9	7.6
13.8	7.5	13.3	8.0	11.3	6.8	7.4	11.7	11.8	7.7
12.6	7.7	13.2	13.9	10.4	12.8	7.6	10.7	10.7	10.9
12.5	11.3	10.7	13.2	8.9	12.9	7.7	9.7	9.7	11.4
11.9	13.4	9.2	13.4	8.8	11.9	7.1	8.5	14.0	14.2

File names   Excel: Tvis05.xls
Minitab: Tvis05.mtp
SPSS: Tvis05.sav
TI-83 Plus and TI-84 Plus/ASCII:
X1 data is stored in Tvis05L1.txt
X2 data is stored in Tvis05L2.txt

06. **Rabies (Two variable independent small samples)**
The following data represent the number of cases of red fox rabies for a random sample of 16 areas in each of two different regions of southern Germany.
Reference: Sayers, B., *Medical Informatics*, Vol. 2, pp. 11-34

X1 = number cases in region 1
10 2 2 5 3 4 3 3 4 0 2 6 4 8 7 4

X2 = number cases in region 2
1 1 2 1 3 9 2 2 4 5 4 2 2 0 0 2

File names Excel: Tvis06.xls
Minitab: Tvis06.mtp
SPSS: Tvis06.sav
TI-83 Plus and TI-84 Plus/ASCII:
X1 data is stored in Tvis06L1.txt
X2 data is stored in Tvis06L2.txt

Copyright © Cengage Learning. All rights reserved.

**07. Weight of Football Players Versus Weight of Basketball Players
(Two variable independent small samples)**
The following data represent weights in pounds of 21 randomly selected pro football players, and 19 randomly selected pro basketball players.
Reference: *Sports Encyclopedia of Pro Football* and *Official NBA Basketball Encyclopedia*

$X1$ = weights (lb.) of pro football players
245	262	255	251	244	276	240	265	257	252	282
256	250	264	270	275	245	275	253	265	270	

$X2$ = weights (lb.) of pro basketball players
205	200	220	210	191	215	221	216	228	207
225	208	195	191	207	196	181	193	201	

File names   Excel: Tvis07.xls
                     Minitab: Tvis07.mtp
                     SPSS: Tvis07.sav
                     TI-83 Plus and TI-84 Plus/ASCII:
                               X1 data is stored in Tvis07L1.txt
                               X2 data is stored in Tvis07L2.txt

08. Birth Rate (Two variable independent small samples)
The following data represent birth rates (per 1000 residential population) for independent random samples of counties in California and Maine.
Reference: *County and City Data Book,* 12th edition, U.S. Dept. of Commerce

$X1$ = birth rate in California counties
14.1	18.7	20.4	20.7	16.0	12.5	12.9	9.6	17.6
18.1	14.1	16.6	15.1	18.5	23.6	19.9	19.6	14.9
17.7	17.8	19.1	22.1	15.6				

$X2$ = birth rate in Maine counties
15.1	14.0	13.3	13.8	13.5	14.2	14.7	11.8	13.5	13.8
16.5	13.8	13.2	12.5	14.8	14.1	13.6	13.9	15.8	

File names   Excel: Tvis08.xls
                     Minitab: Tvis08.mtp
                     SPSS: Tvis08.sav
                     TI-83 Plus and TI-84 Plus/ASCII:
                               X1 data is stored in Tvis08L1.txt
                               X2 data is stored in Tvis08L2.txt

Copyright © Cengage Learning. All rights reserved.

09. Death Rate (Two variable independent small samples)
The following data represent death rates (per 1000 resident population) for independent random samples of counties in Alaska and Texas.
Reference: *County and City Data Book,* 12th edition, U.S. Dept. of Commerce

$X1$ = death rate in Alaska counties
1.4   4.2   7.3   4.8   3.2   3.4   5.1   5.4
6.7   3.3   1.9   8.3   3.1   6.0   4.5   2.5

$X2$ = death rate in Texas counties
7.2   5.8   10.5   6.6   6.9   9.5   8.6   5.9   9.1
5.4   8.8   6.1   9.5   9.6   7.8   10.2   5.6   8.6

File names	Excel: Tvis09.xls

Minitab: Tvis09.mtp
SPSS: Tvis09.sav
TI-83 Plus and TI-84 Plus/ASCII:
        X1 data is stored in Tvis09L1.txt
        X2 data is stored in Tvis09L2.txt

10. Pickup Trucks (Two variable independent small samples)
The following data represent the retail prices (in thousands of dollars) for independent random samples of models of pickup trucks.
Reference: *Consumer Guide,* Vol. 681

$X1$ = prices for different GMC Sierra 1500 models
17.4   23.3   29.2   19.2   17.6   19.2   23.6   19.5   22.2
24.0   26.4   23.7   29.4   23.7   26.7   24.0   24.9

$X2$ = prices for different Chevrolet Silverado 1500 models
17.5   23.7   20.8   22.5   24.3   26.7   24.5   17.8
29.4   29.7   20.1   21.1   22.1   24.2   27.4   28.1

File names   Excel: Tvis10.xls
Minitab: Tvis10.mtp
SPSS: Tvis10.sav
TI-83 Plus and TI-84 Plus/ASCII:
        X1 data is stored in Tvis10L1.txt
        X2 data is stored in Tvis10L2.txt

Copyright © Cengage Learning. All rights reserved.

TWO VARIABLE DEPENDENT SAMPLES

File name prefix: Tvds followed by the number of the data file

01. Average Faculty Salary, Male versus Female (Two variable dependent samples)

In the following data pairs, A = average salaries for males ($1000/yr) and B = average salaries for females ($1000/yr) for assistant professors at the same college or university. A random sample of 22 U.S. colleges and universities was used.

Reference: *Academe, Bulletin of the American Association of University Professors*

A: 34.5   30.5   35.1   35.7   31.5   34.4   32.1   30.7   33.7   35.3
B: 33.9   31.2   35.0   34.2   32.4   34.1   32.7   29.9   31.2   35.5

A: 30.7   34.2   39.6   30.5   33.8   31.7   32.8   38.5   40.5   25.3
B: 30.2   34.8   38.7   30.0   33.8   32.4   31.7   38.9   41.2   25.5

A: 28.6   35.8
B: 28.0   35.1

File names   Excel: Tvds01.xls
                     Minitab: Tvds01.mtp
                     SPSS: Tvds01.sav
                     TI-83 Plus and TI-84 Plus/ASCII:
                                 X1 data is stored in Tvds01L1.txt
                                 X2 data is stored in Tvds01L2.txt

02. Unemployment for College Graduates Versus High School Only
(Two variable dependent samples)

In the following data pairs, A = percent unemployment for college graduates and B = percent unemployment for high school only graduates. The data are paired by year.

Reference: *Statistical Abstract of the United States*

A: 2.8   2.2   2.2   1.7   2.3   2.3   2.4   2.7   3.5   3.0   1.9   2.5
B: 5.9   4.9   4.8   5.4   6.3   6.9   6.9   7.2   10.0   8.5   5.1   6.9

File names   Excel: Tvds02.xls
                     Minitab: Tvds02.mtp
                     SPSS: Tvds02.sav
                     TI-83 Plus and TI-84 Plus/ASCII:
                                 X1 data is stored in Tvds02L1.txt
                                 X2 data is stored in Tvds02L2.txt

Copyright © Cengage Learning. All rights reserved.

03. **Number of Navajo Hogans versus Modern Houses (Two variable dependent samples)**
In the following data pairs, A = number of traditional Navajo hogans in a given district and B = number of modern houses in a given district. The data are paired by district of the Navajo reservation. A random sample of 8 districts was used.
Reference: *Navajo Architecture, Forms, History, Distributions* by S.C. Jett and V.E. Spencer, Univ. of Arizona Press

A:	13	14	46	32	15	47	17	18
B:	18	16	68	9	11	28	50	50

File names   Excel: Tvds03.xls
Minitab: Tvds03.mtp
SPSS: Tvds03.sav
TI-83 Plus and TI-84 Plus/ASCII:
     X1 data is stored in Tvds03L1.txt
     X2 data is stored in Tvds03L2.txt

04. **Temperatures in Miami versus Honolulu (Two variable dependent samples)**
In the following data pairs, A = average monthly temperature in Miami and B = average monthly temperature in Honolulu. The data are paired by month.
Reference: U.S. Department of Commerce Environmental Data Service

A:	67.5	68.0	71.3	74.9	78.0	80.9	82.2	82.7	81.6	77.8	72.3	68.5
B:	74.4	72.6	73.3	74.7	76.2	78.0	79.1	79.8	79.5	78.4	76.1	73.7

File names   Excel: Tvds04.xls
Minitab: Tvds04.mtp
SPSS: Tvds04.sav
TI-83 Plus and TI-84 Plus/ASCII:
     X1 data is stored in Tvds04L1.txt
     X2 data is stored in Tvds04L2.txt

05. **January/February Ozone Column (Two variable dependent samples)**
In the following pairs, the data represent the thickness of the ozone column in Dobson units: one milli-centimeter of ozone at standard temperature and pressure.
     A = monthly mean thickness in January
     B = monthly mean thickness in February
The data are paired by year for a random sample of 15 years.
Reference: Laboratorium für Atmosphärensphysik, Switzerland

A:	360	324	377	336	383	361	369	349
B:	365	325	359	352	397	351	367	397

A:	301	354	344	329	337	387	378
B:	335	338	349	393	370	400	411

File names   Excel: Tvds05.xls
Minitab: Tvds05.mtp
SPSS: Tvds05.sav
TI-83 Plus and TI-84 Plus/ASCII:
     X1 data is stored in Tvds05L1.txt
     X2 data is stored in Tvds05L2.txt

Copyright © Cengage Learning. All rights reserved.

06.   Birth Rate/Death Rate (Two variable dependent samples)

In the following data pairs, A = birth rate (per 1000 resident population) and B = death rate (per 1000 resident population). The data are paired by county in Iowa.

Reference: *County and City Data Book*, 12th edition, U.S. Dept. of Commerce

A:	12.7	13.4	12.8	12.1	11.6	11.1	14.2
B:	9.8	14.5	10.7	14.2	13.0	12.9	10.9

A:	12.5	12.3	13.1	15.8	10.3	12.7	11.1
B:	14.1	13.6	9.1	10.2	17.9	11.8	7.0

File names   Excel: Tvds06.xls
                    Minitab: Tvds06.mtp
                    SPSS: Tvds06.sav
                    TI-83 Plus and TI-84 Plus/ASCII:
                              X1 data is stored in Tvds06L1.txt
                              X2 data is stored in Tvds06L2.txt

07.   Democrat/Republican (Two variable dependent samples)

In the following data pairs A = percentage of voters who voted Democrat and B = percentage of voters who voted Republican in a recent national election. The data are paired by county in Indiana.

Reference: *County and City Data Book*, 12th edition, U.S. Dept. of Commerce

A:	42.2	34.5	44.0	34.1	41.8	40.7	36.4	43.3	39.5
B:	35.4	45.8	39.4	40.0	39.2	40.2	44.7	37.3	40.8

A:	35.4	44.1	41.0	42.8	40.8	36.4	40.6	37.4
B:	39.3	36.8	35.5	33.2	38.3	47.7	41.1	38.5

File names   Excel: Tvds07.xls
                    Minitab: Tvds07.mtp
                    SPSS: Tvds07.sav
                    TI-83 Plus and TI-84 Plus/ASCII:
                              X1 data is stored in Tvds07L1.txt
                              X2 data is stored in Tvds07L2.txt

08.   Santiago Pueblo Pottery (Two variable dependent samples)

In the following data, A = percentage of utility pottery and B = percentage of ceremonial pottery found at the Santiago Pueblo archaeological site. The data are paired by location of discovery.

Reference: Laboratory of Anthropology, Notes 475, Santa Fe, New Mexico

A:	41.4	49.6	55.6	49.5	43.0	54.6	46.8	51.1	43.2	41.4
B:	58.6	50.4	44.4	59.5	57.0	45.4	53.2	48.9	56.8	58.6

File names   Excel: Tvds08.xls
                    Minitab: Tvds08.mtp
                    SPSS: Tvds08.sav
                    TI-83 Plus and TI-84 Plus/ASCII:
                              X1 data is stored in Tvds08L1.txt
                              X2 data is stored in Tvds08L2.txt

Copyright © Cengage Learning. All rights reserved.

09. Poverty Level (Two variable dependent samples)

In the following data pairs, A = percentage of population below poverty level in 1998 and B = percentage of population below poverty level in 1990. The data are grouped by state and District of Columbia.

Reference: *Statistical Abstract of the United States*, 120th edition

A: 14.5   9.4   16.6   14.8   15.4   9.2   9.5   10.3   22.3   13.1
B: 19.2   11.4   13.7   19.6   13.9   13.7   6.0   6.9   21.1   14.4

A: 13.6   10.9   13.0   10.1   9.4   9.1   9.6   13.5   19.1   10.4
B: 15.8   11.0   14.9   13.7   13.0   10.4   10.3   17.3   23.6   13.1

A: 7.2   8.7   11.0   10.4   17.6   9.8   16.6   12.3   10.6   9.8
B: 9.9   10.7   14.3   12.0   25.7   13.4   16.3   10.3   9.8   6.3

A: 8.6   20.4   16.7   14.0   15.1   11.2   14.1   15.0   11.2   11.6
B: 9.2   20.9   14.3   13.0   13.7   11.5   15.6   9.2   11.0   7.5

A: 13.7   10.8   13.4   15.1   9.0   9.9   8.8   8.9   17.8   8.8   10.6
B: 16.2   13.3   16.9   15.9   8.2   10.9   11.1   8.9   18.1   9.3   11.0

File names   Excel: Tvds09.xls
                      Minitab: Tvds09.mtp
                      SPSS: Tvds09.sav
                      TI-83 Plus and TI-84 Plus/ASCII:
                              X1 data is stored in Tvds09L1.txt
                              X2 data is stored in Tvds09L2.txt

10. Cost of Living Index (Two variable dependent samples)

The following data pairs represent cost of living index for A = grocery items and B = health care. The data are grouped by metropolitan areas.

Reference: *Statistical Abstract of the United States*, 120th edition

A: 96.6   97.5   113.9   88.9   108.3   99.0   97.3   87.5   96.8
B: 91.6   95.9   114.5   93.6   112.7   93.6   99.2   93.2   105.9

A: 102.1   114.5   100.9   100.0   100.7   99.4   117.1   111.3   102.2
B: 110.8   127.0   91.5   100.5   104.9   104.8   124.1   124.6   109.1

A: 95.3   91.1   95.7   87.5   91.8   97.9   97.4   102.1   94.0
B: 98.7   95.8   99.7   93.2   100.7   96.0   99.6   98.4   94.0

A: 115.7   118.3   101.9   88.9   100.7   99.8   101.3   104.8   100.9
B: 121.2   122.4   110.8   81.2   104.8   109.9   103.5   113.6   94.6

A: 102.7   98.1   105.3   97.2   105.2   108.1   110.5   99.3   99.7
B: 109.8   97.6   109.8   107.4   97.7   124.2   110.9   106.8   94.8

File names   Excel: Tvds10.xls
                      Minitab: Tvds10.mtp
                      SPSS: Tvds10.sav
                      TI-83 Plus and TI-84 Plus/ASCII:
                              X1 data is stored in Tvds10L1.txt
                              X2 data is stored in Tvds10L2.txt

Copyright © Cengage Learning. All rights reserved.

SIMPLE LINEAR REGRESSION

File name prefix: Slr followed by the number of the data file

01. List Price versus Best Price for a New GMC Pickup Truck (Simple Linear Regression)
In the following data pairs, X = list price (in $1000) for a GMC pickup truck and Y = best price (in $1000) for a GMC pickup truck.
Reference: *Consumer's Digest*

X:	12.4	14.3	14.5	14.9	16.1	16.9	16.5	15.4	17.0	17.9
Y:	11.2	12.5	12.7	13.1	14.1	14.8	14.4	13.4	14.9	15.6

X:	18.8	20.3	22.4	19.4	15.5	16.7	17.3	18.4	19.2	17.4
Y:	16.4	17.7	19.6	16.9	14.0	14.6	15.1	16.1	16.8	15.2

X:	19.5	19.7	21.2
Y:	17.0	17.2	18.6

File names   Excel: Slr01.xls
          Minitab: Slr01.mtp
          SPSS: Slr01.sav
          TI-83 Plus and TI-84 Plus/ASCII:
               X1 data is stored in Slr01L1.txt
               X2 data is stored in Slr01L2.txt

02. Cricket Chirps versus Temperature (Simple Linear Regression)
In the following data pairs, X = chirps/sec for the striped ground cricket and Y = temperature in degrees Fahrenheit.
Reference: *The Song of Insects* by Dr. G.W. Pierce, Harvard College Press

X:	20.0	16.0	19.8	18.4	17.1	15.5	14.7	17.1
Y:	88.6	71.6	93.3	84.3	80.6	75.2	69.7	82.0

X:	15.4	16.2	15.0	17.2	16.0	17.0	14.4
Y:	69.4	83.3	79.6	82.6	80.6	83.5	76.3

File names   Excel: Slr02.xls
          Minitab: Slr02.mtp
          SPSS: Slr02.sav
          TI-83 Plus and TI-84 Plus/ASCII:
               X1 data is stored in Slr02L1.txt
               X2 data is stored in Slr02L2.txt

Copyright © Cengage Learning. All rights reserved.

03. Diameter of Sand Granules versus Slope on Beach (Simple Linear Regression)
In the following data pairs, X = median diameter (mm) of granules of sand and Y = gradient of beach slope in degrees. The data are for naturally occurring ocean beaches.
Reference: *Physical Geography* by A.M. King, Oxford Press, England

X:	0.170	0.190	0.220	0.235	0.235	0.300	0.350	0.420	0.850
Y:	0.630	0.700	0.820	0.880	1.150	1.500	4.400	7.300	11.300

File names   Excel: Slr03.xls
Minitab: Slr03.mtp
SPSS: Slr03.sav
TI-83 Plus and TI-84 Plus/ASCII:
       X1 data is stored in Slr03L1.txt
       X2 data is stored in Slr03L2.txt

04. National Unemployment Male versus Female (Simple Linear Regression)
In the following data pairs, X = national unemployment rate for adult males and Y = national unemployment rate for adult females.
Reference: *Statistical Abstract of the United States*

X:	2.9	6.7	4.9	7.9	9.8	6.9	6.1	6.2	6.0	5.1	4.7	4.4	5.8
Y:	4.0	7.4	5.0	7.2	7.9	6.1	6.0	5.8	5.2	4.2	4.0	4.4	5.2

File names   Excel: Slr04.xls
Minitab: Slr04.mtp
SPSS: Slr04.sav
TI-83 Plus and TI-84 Plus/ASCII:
       X1 data is stored in Slr04L1.txt
       X2 data is stored in Slr04L2.txt

05. Fire and Theft in Chicago (Simple Linear Regression)
In the following data pairs, X = fires per 1000 housing units and Y = thefts per 1000 population within the same zip code in the Chicago metro area.
Reference: U.S. Commission on Civil Rights

X:	6.2	9.5	10.5	7.7	8.6	34.1	11.0	6.9	7.3	15.1
Y:	29	44	36	37	53	68	75	18	31	25

X:	29.1	2.2	5.7	2.0	2.5	4.0	5.4	2.2	7.2	15.1
Y:	34	14	11	11	22	16	27	9	29	30

X:	16.5	18.4	36.2	39.7	18.5	23.3	12.2	5.6	21.8	21.6
Y:	40	32	41	147	22	29	46	23	4	31

X:	9.0	3.6	5.0	28.6	17.4	11.3	3.4	11.9	10.5	10.7
Y:	39	15	32	27	32	34	17	46	42	43

X:	10.8	4.8
Y:	34	19

File names   Excel: Slr05.xls
Minitab: Slr05.mtp
SPSS: Slr05.sav
TI-83 Plus and TI-84 Plus/ASCII:
       X1 data is stored in Slr05L1.txt
       X2 data is stored in Slr05L2.txt

Copyright © Cengage Learning. All rights reserved.

06. Auto Insurance in Sweden (Simple Linear Regression)

In the following data pairs, X = number of claims and Y = total payment for all the claims in thousands of Swedish Kronor for geographical zones in Sweden.

Reference: Swedish Committee on Analysis of Risk Premium in Motor Insurance

X:	108	19	13	124	40	57	23	14	45	10
Y:	392.5	46.2	15.7	422.2	119.4	170.9	56.9	77.5	214.0	65.3

X:	5	48	11	23	7	2	24	6	3	23
Y:	20.9	248.1	23.5	39.6	48.8	6.6	134.9	50.9	4.4	113.0

X:	6	9	9	3	29	7	4	20	7	4
Y:	14.8	48.7	52.1	13.2	103.9	77.5	11.8	98.1	27.9	38.1

X:	0	25	6	5	22	11	61	12	4	16
Y:	0.0	69.2	14.6	40.3	161.5	57.2	217.6	58.1	12.6	59.6

X:	13	60	41	37	55	41	11	27	8	3
Y:	89.9	202.4	181.3	152.8	162.8	73.4	21.3	92.6	76.1	39.9

X:	17	13	13	15	8	29	30	24	9	31
Y:	142.1	93.0	31.9	32.1	55.6	133.3	194.5	137.9	87.4	209.8

X:	14	53	26
Y:	95.5	244.6	187.5

File names	Excel: Slr06.xls
	Minitab: Slr06.mtp
	SPSS: Slr06.sav
	TI-83 Plus and TI-84 Plus/ASCII:
	X1 data is stored in Slr06L1.txt
	X2 data is stored in Slr06L2.txt

Copyright © Cengage Learning. All rights reserved.

07. **Gray Kangaroos (Simple Linear Regression)**

In the following data pairs, X = nasal length (mm × 10) and Y = nasal width (mm × 10) for a male gray kangaroo from a random sample of such animals.

Reference: *Australian Journal of Zoology*, Vol. 28, pp. 607-613

X:	609	629	620	564	645	493	606	660	630	672
Y:	241	222	233	207	247	189	226	240	215	231

X:	778	616	727	810	778	823	755	710	701	803
Y:	263	220	271	284	279	272	268	278	238	255

X:	855	838	830	864	635	565	562	580	596	597
Y:	308	281	288	306	236	204	216	225	220	219

X:	636	559	615	740	677	675	629	692	710	730
Y:	201	213	228	234	237	217	211	238	221	281

X:	763	686	717	737	816
Y:	292	251	231	275	275

File names   Excel: Slr07.xls
                  Minitab: Slr07.mtp
                  SPSS: Slr07.sav
                  TI-83 Plus and TI-84 Plus/ASCII:
                          X1 data is stored in Slr07L1.txt
                          X2 data is stored in Slr07L2.txt

08. **Pressure and Weight in Cryogenic Flow Meters (Simple Linear Regression)**

In the following data pairs, X = pressure (lb./sq. in.) of liquid nitrogen and Y = weight in pounds of liquid nitrogen passing through a flow meter each second.

Reference: *Technometrics*, Vol. 19, pp. 353-379

X:	75.1	74.3	88.7	114.6	98.5	112.0	114.8	62.2	107.0
Y:	577.8	577.0	570.9	578.6	572.4	411.2	531.7	563.9	406.7

X:	90.5	73.8	115.8	99.4	93.0	73.9	65.7	66.2	77.9
Y:	507.1	496.4	505.2	506.4	510.2	503.9	506.2	506.3	510.2

X:	109.8	105.4	88.6	89.6	73.8	101.3	120.0	75.9	76.2
Y:	508.6	510.9	505.4	512.8	502.8	493.0	510.8	512.8	513.4

X:	81.9	84.3	98.0
Y:	510.0	504.3	522.0

File names   Excel: Slr08.xls
                  Minitab: Slr08.mtp
                  SPSS: Slr08.sav
                  TI-83 Plus and TI-84 Plus/ASCII:
                          X1 data is stored in Slr08L1.txt
                          X2 data is stored in Slr08L2.txt

Copyright © Cengage Learning. All rights reserved.

09.   Ground Water Survey (Simple Linear Regression)

In the following data pairs, X = pH of well water and Y = bicarbonate (parts per million) of well water. The data are for well water from a random sample of wells in Northwest Texas.
Reference: Nichols, C.E. and Kane, V.E., Union Carbide Technical Report K/UR-1

X:	7.6	7.1	8.2	7.5	7.4	7.8	7.3	8.0	7.1	7.5
Y:	157	174	175	188	171	143	217	190	142	190

X:	8.1	7.0	7.3	7.8	7.3	8.0	8.5	7.1	8.2	7.9
Y:	215	199	262	105	121	81	82	210	202	155

X:	7.6	8.8	7.2	7.9	8.1	7.7	8.4	7.4	7.3	8.5
Y:	157	147	133	53	56	113	35	125	76	48

X:	7.8	6.7	7.1	7.3
Y:	147	117	182	87

File names   Excel: Slr09.xls
                    Minitab: Slr09.mtp
                    SPSS: Slr09.sav
                    TI-83 Plus and TI-84 Plus/ASCII:
                           X1 data is stored in Slr09L1.txt
                           X2 data is stored in Slr09L2.txt

10.   *Iris Setosa* (Simple Linear Regression)

In the following data pairs, X = sepal width (cm) and Y = sepal length (cm). The data are for a random sample of the wild flower *iris setosa*.
Reference: Fisher, R.A., *Ann. Eugenics,* Vol. 7, Part II, pp. 179-188

X:	3.5	3.0	3.2	3.1	3.6	3.9	3.4	3.4	2.9	3.1
Y:	5.1	4.9	4.7	4.6	5.0	5.4	4.6	5.0	4.4	4.9

X:	3.7	3.4	3.0	4.0	4.4	3.9	3.5	3.8	3.8	3.4
Y:	5.4	4.8	4.3	5.8	5.7	5.4	5.1	5.7	5.1	5.4

X:	3.7	3.6	3.3	3.4	3.0	3.4	3.5	3.4	3.2	3.1
Y:	5.1	4.6	5.1	4.8	5.0	5.0	5.2	5.2	4.7	4.8

X:	3.4	4.1	4.2	3.1	3.2	3.5	3.6	3.0	3.4	3.5
Y:	5.4	5.2	5.5	4.9	5.0	5.5	4.9	4.4	5.1	5.0

X:	2.3	3.2	3.5	3.8	3.0	3.8	3.7	3.3
Y:	4.5	4.4	5.0	5.1	4.8	4.6	5.3	5.0

File names   Excel: Slr10.xls
                    Minitab: Slr10.mtp
                    SPSS: Slr10.sav
                    TI-83 Plus and TI-84 Plus/ASCII:
                           X1 data is stored in Slr10L1.txt
                           X2 data is stored in Slr10L2.txt

Copyright © Cengage Learning. All rights reserved.

11. **Pizza Franchise (Simple Linear Regression)**
    In the following data pairs, X = annual franchise fee ($1000) and Y = start-up cost ($1000) for a pizza franchise.
    Reference: *Business Opportunities Handbook*

X:	25.0	8.5	35.0	15.0	10.0	30.0	10.0	50.0	17.5	16.0
Y:	125	80	330	58	110	338	30	175	120	135

X:	18.5	7.0	8.0	15.0	5.0	15.0	12.0	15.0	28.0	20.0
Y:	97	50	55	40	35	45	75	33	55	90

X:	20.0	15.0	20.0	25.0	20.0	3.5	35.0	25.0	8.5	10.0
Y:	85	125	150	120	95	30	400	148	135	45

X:	10.0	25.0
Y:	87	150

File names	Excel: Slr11.xls
	Minitab: Slr11.mtp
	SPSS: Slr11.sav
	TI-83 Plus and TI-84 Plus/ASCII:
	X1 data is stored in Slr11L1.txt
	X2 data is stored in Slr11L2.txt

12. **Prehistoric Pueblos (Simple Linear Regression)**
    In the following data pairs, X = estimated year of initial occupation and Y = estimated year of end of occupation. The data are for each prehistoric pueblo in a random sample of such pueblos in Utah, Arizona, and Nevada.
    Reference: *Prehistoric Pueblo World*, by A. Adler, Univ. of Arizona Press

X:	1000	1125	1087	1070	1100	1150	1250	1150	1100
Y:	1050	1150	1213	1275	1300	1300	1400	1400	1250

X:	1350	1275	1375	1175	1200	1175	1300	1260	1330
Y:	1830	1350	1450	1300	1300	1275	1375	1285	1400

X:	1325	1200	1225	1090	1075	1080	1080	1180	1225
Y:	1400	1285	1275	1135	1250	1275	1150	1250	1275

X:	1175	1250	1250	750	1125	700	900	900	850
Y:	1225	1280	1300	1250	1175	1300	1250	1300	1200

File names	Excel: Slr12.xls
	Minitab: Slr12.mtp
	SPSS: Slr12.sav
	TI-83 Plus and TI-84 Plus/ASCII:
	X1 data is stored in Slr12L1.txt
	X2 data is stored in Slr12L2.txt

Copyright © Cengage Learning. All rights reserved.